Unveiling Creation:

Eight is the Key

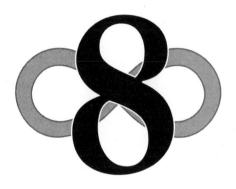

by

Nirmal D.C. Pugh

&

Derek C. Pugh, PhD

Sunstar
PUBLISHING LTD.

UNVEILING CREATION:
Eight is the Key
by Nirmal D.C. Pugh
and Derek C. Pugh, PhD

©1999, United States Copyright
Sunstar Publishing, Ltd.
204 North Court Street
Fairfield, Iowa 52556

Cover Design: Amanda Collett
Editor: Elizabeth Pasco
Page Layout & Design: Sharon A. Dunn

LCCN: 97-061869
ISBN: 1-887472-43-6
Printed in the U.S.A.

Readers interested in obtaining further information on the subject matter of
this book are invited to correspond with:

The Secretary, Sunstar Publishing, Ltd.
204 North Court Street, Fairfield, Iowa 52556

For more Sunstar Books, visit: *http://www.newagepage.com*

NIRMAL D.C. PUGH

Nirmal Pugh, a native of Canada, has been an avid practitioner of meditation since he was four years old. At the age of twelve he moved to Fairfield, Iowa, to spend the following twelve years at Maharishi International University (MIU) assimilating an extensive understanding of ancient Vedic knowledge and spirituality. Also during this period he learned both the western and traditional Indian bamboo flute, resulting in a gold medal at the 1989 state competition in Iowa.

Complementing this background, Mr. Pugh graduated summa cum laude in 1994 as class valedictorian with a BS degree in biochemistry and biology from MIU. He has published several peer-review articles and presented research at national and international conferences on the neurochemical effects of meditation, consciousness and approaches of Ayurveda. During the past year he was invited to join the Rho Chi Pharmaceutical Honor Society and is currently pursuing a PhD degree at the University of Mississippi in pharmacognosy – the discovery and development of therapeutic agents from natural products.

DEREK C. PUGH

Derek Pugh's quest for Freedom and Truth began in London. By age seven he was exploring English hills. At ten he was contemplating the mysteries of time and space. During World War II he graduated in math and physics from Rutlish School, Merton, and entered London University. His first spiritual awakening came in 1945 while hiking Scottish mountains. His peak athletic performance was winning the 1950 European Championships 400 metres. He then entered Paris University, recovered from polio and gained his PhD in geology. His personal quest continued into the Canadian Rockies, led him to meditation and culminated in India for three wonderful months with Maharishi Mahesh Yogi.

In 1984, after thirty years of research with Geological Survey of Canada, Derek retired with the accolade "world class geologist." Since then he has been a researcher in consciousness. His goal is highest human development and fulfillment of long-term global dreams.

TABLE OF CONTENTS

LIST OF FIGURES

LIST OF TABLES

ACKNOWLEDGMENTS

"We have inherited from our forefathers the keen longing for unified, all-embracing knowledge. The very name given to the highest institutions of learning reminds us, that from antiquity and throughout many centuries the universal aspect has been the only one to be given full credit."

– *Erwin Schrodinger,* 20th century physicist
Preface to *"What is Life?"* (1945)

This book was made possible by the loving support and encouragement of the following people:

- Maharishi Mahesh Yogi and his master Swami Brahmananda Saraswati, Jagadguru, Bhagwan Shankaracharya of Jyotir Math, Himalayas, for their divine inspiration and exceptionally clear restoration of the ancient Vedic wisdom of life and our universe;

- Steven Kelley and Sunil Rawal for invaluable information on superstring theories in physics;

- Vera Gartley for general knowledge regarding the color theories of art;

- Brad Stewart for valuable discussions and references on Sacred Geometry and Ancient wisdom;

- Douglass White for expert advice on the I Ching and the Mayan Calendar;

- Evan Finkelstein for critical references and guidance in understanding the Kabbalah;

- Vikas Narula for technical assistance with the diagrams and figures; and

- **Friends and family** for numerous suggestions and deep discussions.

We would also like to give a special thanks to **Tricia Pierre** and **Kelly Sterling** for their inspiration, support, encouragement and proofreading. Thank you all. We hope this book reflects the deepest wisdom and the secret teachings of the ages for the future happiness of mankind.

FOREWORD

By Dr. David Frawley
(Vamadeva Shastri)
April 1998

The Rig Veda is one of the greatest repositories of secret knowledge, being the very basis of the great spiritual traditions of India, including Yoga and Vedanta. It is also the oldest book in the world. Recent revised estimates based upon the latest archeologist research suggest an antiquity of more than five thousand years for this important text. This is based upon three main points.

First, the Rig Veda describes the Sarasvati river as its central location, the motherland of its culture. LandSat photography, along with various excavations of the river bed and the settlements along it, reveal that this river which flowed east of the Indus was once the largest in India. It dried up around 2000 BC after a series of droughts and earthquakes over several centuries that changed its course several times. The Rig Veda therefore must be older than the terminal period of the river to have developed around it and to know so intimately its terrain.

Second, archeology reveals that the Sarasvati river, not the Indus, was the main river of the massive urban culture previously called the Harappan or Indus Valley of the third millennium BC, including the famous site of Mohenjo-daro. Presently five sites have been identified of the same size as Mohenjo-daro and Harappa, the other three, Rakhigeri, Ganweriwala and Dholarvira, being on the Sarasvati and its tributaries. In addition recent excavation has uncovered Vedic or Aryan traits in Harappan and pre-Harappan sites including many fire altars and horse remains, showing the same type of culture in these ruins as described in Vedic and post-Vedic texts.

Third, archeo-astronomy reveals that Vedic texts were aware of astronomical positions of the Pleiades (Krittika) vernal equinox of 2500 BC

and earlier, reflecting the era of the Orion (Mrigashirsha) equinox
of 4000 BC as well as points perhaps anterior to this. Putting these
together a new multidisciplinary approach reveals the antiquity of Vedic
civilization in India and from there influencing the entire world. What
is unique about this civilization is its spiritual and yogic basis which
provides a counterpoint to the materialistic culture that has come out
of Europe and the West.

This means that the Rig Veda represents the heritage of one of the great
civilizations of the ancient world and the one that has best preserved its
continuity through history. The Rig Veda is also well known as the oldest
book in an Indo-European language and the repository of the spiritual
and religious knowledge of the early Indo-Europeans. The ancient
Greeks, Romans, Germans, Kelts and Slavs practiced a religion similar
to the Vedic, worshipped Gods and Goddesses of similar names and
qualities, and spoke a similar language. Therefore the Vedas reflect our
older European heritage and are relevant for anyone seeking those out
as well. We could say that the Vedas contain much of the secret of the
origins of civilization and, particularly, the spiritual heritage of humanity.
In fact, according to the Vedic tradition, Vedic knowledge is inherent in
the cosmic mind. The Vedas project not only the earlier spiritual knowledge of humanity but that of all intelligent life in the universe.

The Rig Veda is first of all a spiritual or religious book. Its main concern
is Dharma, which refers to universal law. These laws in the Vedic view
are also the laws of consciousness and the laws of our own deeper
immortal nature. Through understanding Dharma we are able to live
in harmony with the universe and to discover the entire universe within
us as our own deeper Self.

The Rig Veda consists of various mantras or words of spiritual power
and insight that were said to have been cognized by the rishis or sages
directly from the Divine or Cosmic Mind. It contains one thousand
hymns and ten thousand verses in a very compact and subtle language.
The tradition is that the secrets of both cosmogenesis and higher human
evolution are hidden in its teachings and in the structure of its language,
which contains the very power of creation.

However, the Vedas say over and over again that the Gods, which also
means the sages, prefer what is cryptic or secretive (paroksha priya hi
deva). They don't speak in an obvious language so as to protect their

wisdom from the ignorant. The Vedic texts therefore are sealed in a secret language and were originally only accessible to the initiated. Lacking this spiritual key of initiation the modern scholars who have examined Vedic texts from a purely intellectual perspective along with their own social, political and cultural biases have not surprisingly failed to find much deep meaning in the text at all, little more than primitive poetry and superstition. Yet any deep field of knowledge requires such a secret key. What would an outsider think some thousands of years from now of modern physics and its talk of quarks and quasars, and particles with strangeness, if he did not have the key to the science behind it?

Fortunately the great Yogis of modern India have given us a different view and provide us a way of entrance to the Vedic world. Swami Dayananda Saraswati, Sri Aurobindo, Yogi Ganapati Muni and his disciple Brahmarshi Daivarata have provided us these keys. Their work, however, is either hard to find or much of it remains in Sanskrit only. So fortunately again the great teacher Maharishi Mahesh Yogi has made such an approach accessible to the English speaking world and promoted this view of the Vedas world wide. In the footsteps of these great teachers individuals in the West have begun looking at these deeper secrets of the Vedas and correlating them with other forms of spiritual knowledge, with modern science, and adding their own unique insights.

It is in this background that I am happy and honored to introduce *Unveiling Creation: Eight is the Key* by Nirmal D.C. Pugh and Derek C. Pugh. The book is one of the most insightful new presentations of Vedic knowledge in the Western world to come out in recent years. It hopefully indicates a greater trend of study that many other researchers will follow in the future. The authors weave a marvelous correlation of our most ancient spiritual wisdom with the most recent findings of modern physics that is similarly taking us beyond the limits of space and time. They help bridge the gap between science and spirit and between the primal and the futuristic toward a new world vision and a new world order for the coming planetary age.

Number in the Vedic science is one of the faces of the Gods or the cosmic powers, through which the workings of the universe are revealed. The Vedas delight in number. The Yajur Veda first introduces the decimal system in a comprehensive way noting numbers from one to ten to one hundred to 1,000,000,000,000. Number is not just an abstraction but a reflection of the essence of the universe and the key to its structure. Nor does number exist apart from the universal forces of life and intelligence.

The authors show the numerical code of the universe as revealed both by modern science and by Vedic science. They show that it is also a musical code and how it is reflected in the colors, energies and vital forces active around us. In this regard Vedic numbers appear in the meters and syllable counts of various hymns, and as a key to the music of the Sama Veda. They are reflected in the elements and the biological humors that structure physical and biological life. Putting this system together the authors help us uncover the greater spiritual, sacred or universal science that we are all seeking beyond our various efforts to learn the nature of life and consciousness.

Most notably, the authors show how this Vedic code is reflected in other ancient systems of spiritual knowledge including the Greek, the Persian, the Chinese and the Mayan. After all if this is a universal code, any person of insight and higher consciousness should be able to find it. Yet it is also possible that these other ancient cultures received this knowledge from contact with India, particularly the Persians who were originally part of Vedic civilization.

What is perhaps most amazing is how simple this code ultimately is, a matter of a common integer like the number eight. The highest knowledge returns us to simplicity, which is to take us back to the One of which all multiplicity is but a display like the colors of the rainbow or the tail of a peacock. In this way all the complexity and confusion of our lives can be resolved in the unity of consciousness itself.

May the reader take this book as a doorway to Vedic wisdom and through Vedic wisdom to universal understanding!

– Dr. David Frawley
(Vamadeva Shastri)

Founder and Director
American Institute of Vedic Studies
Santa Fe, New Mexico, 87504

Dr. David Frawley is the author of fifteen books and numerous articles on Vedic knowledge, including:

- *Gods, Sages and Kings: Vedic Secrets of Ancient Civilizations;*
- *Wisdom of the Ancient Seers;* and
- *In Search of the Cradle of Civilizations*
  (co-authored with G. Feuerstein and S. Kak).

INTRODUCTION

This book is dedicated to all who are drawn to read it. Our goal has been to touch as many lives as possible with a message of universal truth and understanding. During the course of our research we have found that universal truth has never been the property of any one culture, period of time, or geographical location. Our purpose in writing this book is to provide not only a vivid restoration of the ancient understanding of the exact mechanics governing the universe, but also definitive evidence that this wisdom is found throughout the modern disciplines and ancient civilizations. Because of the broad range of topics covered, everyone will find something of interest. Although many of the concepts we will be discussing are widely known, no one to our knowledge has synthesized these partial values to create a unified perspective that is simple to understand.

In our distant past ancient civilizations have possessed advanced spiritual knowledge about the ultimate nature of our universe. These ancient cultures were guided by spiritual understanding rather than by the materialistic and commercial values that dominate our world today. Our modern idea that man is a physical entity and that our history is merely the development of material tools is incorrect. In reality, our true heritage originates from past civilizations that lived an enlightened way of life guided by holistic knowledge and spiritual fulfillment.

Much of what we know about the ancient world is based on oral traditions, sacred scriptures and archeological remains. From these records it appears that every school of thought and system of knowledge has the same understanding of how creation works. The writings from these civilizations portray knowledge which is not ephemeral, but rather deep wisdom with enormous value to humanity.

Throughout the ages various groups and enlightened sages have restored the complete understanding of the ultimate nature of our universe. Each of these individuals or groups has tried to tell the world about its simplicity and enormous value, but all have been faced with a difficult task. It has been impossible to communicate this wisdom to others verbally because universal knowledge at its very basis is transcendental in its nature. In order to understand the fundamental truths of life it is necessary to have knowledge of the Self which is beyond even our deepest thoughts.

Ancient cultures believed that an omnipresent field of consciousness was responsible for the creation of all aspects of our universe according to definite patterns. This unmanifest field of intelligence was considered to be a storehouse of complete knowledge and a lively domain of all possibilities. According to the Ancients, the key to understanding the secrets of the universe is to have both intellectual knowledge and advanced systems for the development of consciousness. However, because most individuals do not have the personal experience of the inner levels of life, an intellectual understanding of the ultimate reality has been obscure and superficial. For this reason modern scholars who have tried to analyze ancient records from a materialistic angle have failed to understand the real truths and have usually classified them as primitive, unintelligible and mythical.

With the rise of modern science our society has experienced a diminishing credibility and interest in understanding the spiritual nature of the ultimate reality. For the most part world religions today have degraded to a deplorable state where the secret inner teachings of life have been replaced by dogmatic and distorted commentaries. Another great tragedy of modern times is the excessive reductionism, fragmentation and hyper-specialization of knowledge that has resulted in the massive accumulation of details regarding the material nature of our universe. There is a great need for our society to examine this large body of information in search of more holistic and unifying principles. One of the purposes of this book is to demonstrate that hidden within the detailed information in various disciplines of modern science are the primal patterns of intelligence and consciousness – the same fundamental mechanisms understood by ancient cultures thousands of years ago.

In recent years there has been a global upsurge of interest in the spiritual wisdom of the ancient world. Numerous books have been published on these topics, and although many writers have hit upon key ideas, it appears that most individuals still do not have a unified perspective on how everything fits together. Few people comprehend the most primal mechanics of creation or how these fundamental patterns construct our universe. Based on the growing interest in a complete restoration of the ancient wisdom of life we were inspired to write this book. We began several years ago with knowledge derived from the ancient Vedic civilization. The literature from the Vedic age contains the most extensive and largest body of original teachings of any culture in the ancient world. For this reason we felt that the texts of the ancient Vedic culture would be a valuable source for research. It quickly became apparent to us that these writings present a clear account of the exact mechanics responsible for governing creation. We believe that establishing the foundations of universal wisdom will provide the framework to organize and explain all other details.

In this book we have referred to the primal mechanics governing creation as the "Constitution of the Universe."[1] This constitution is literally the template that describes the fundamental patterns of order from which all diversity is constructed. It is a pattern of intelligence which is reflected in the foundational principles of every discipline and thereby provides a common thread that links all systems of ancient and modern teaching. We no longer need to confine our interests to one specialization – now we can grasp the basic wisdom common to all walks of life.

This book is organized into two parts. The first part describes the Constitution of the Universe which governs both the material and subjective aspects of our universe. In the second part we derive from the Constitution of the Universe the exact mechanics for the genesis of the material aspect of our universe. Because of the disparate nature of the topics covered, we have limited the use of technical jargon in order to strike a middle zone so a nonspecialist will be able to grasp the concepts and the expert will not find it an over-simplification. What we are presenting in this book is more than merely "mind games." It is a deep intellectual understanding of how creation works. Just comprehending this intellectual knowledge is enough to change the way we view our surroundings and the way we understand the purpose of life.

This book represents our contribution to the emerging trend of spiritual regeneration in our world at this time. We hope our readers will enjoy reading and re-reading this book as much as we have enjoyed writing and re-writing it. If we can enjoy the truth about life, then certainly we will be able to enjoy living life and be happy. Why else were we born?

---

[1] The phrase "Constitution of the Universe" was originally coined by Maharishi Mahesh Yogi in 1992.

PART A

THE CONSTITUTION
OF THE UNIVERSE

"The rhythm of nature seems to conform to a definite pattern. The infinite number of galaxies in the vast structure of cosmic space seem to move according to a definite plan. The creation, evolution and dissolution of all things seems to follow a definite procedure. Things change, but the incessant change itself seems to have some unchanging basis."

Maharishi Mahesh Yogi, *Bhagavad-Gita,* 1967

Just as a nation's constitution sets forth the most basic laws of legislation in a country, so also a universal constitution exists which determines the most primal laws of nature that govern how creation operates. In this book we refer to this universal constitution as the "Constitution of the Universe." This phrase represents the most fundamental structure of intelligence and order from which the entire universe is generated and governed. The chapters making up the first part of this book provide an intellectual understanding of the Constitution of the Universe and locate its presence throughout creation.

Our description of the Constitution of the Universe is based on the literature of the Vedic civilization of ancient India. The insights contained in these age-old texts provide one of the clearest accounts of how creation works. Although there are also numerous other ancient cultures which had the same understanding, the wisdom from these other systems has been hidden or poorly understood. Our purpose in this first part of the book is two-fold. First, we present evidence that the structure of the Constitution of the Universe is hidden within the fundamental aspects of the modern disciplines. Second, we demonstrate that all major civilizations of the past possessed an identical understanding of the primal mechanics inherent in the Constitution of the Universe. Based on this research we feel that there is enough evidence to establish a complete understanding of the Constitution of the Universe which can be used as a central reference for evaluating and restoring other systems of knowledge.

In brief, the organization or structure of the Constitution of the Universe consists of two levels: the first level consists of eight fundamental branches which emerge from a unified source; in the second stage, each of these eight basic units are elaborated eight times to create sixty-four aspects. At both levels there is a three-fold interpretation, or three different ways of understanding, the eight and the sixty-four divisions. The names for the eight and sixty-four divisions in the Constitution of the Universe differ according to the language of the discipline or nature of the ancient teaching. Although the structure of the Constitution of the Universe may appear straightforward, we need to remind ourselves that universal truths are simple.

In this first part of the book we suggest reading chapter one as it introduces the details of the Constitution of the Universe as understood by the Vedic civilization of ancient India. Since the subsequent chapters furnish evidence that this same template also occurs in the primal teachings of other ancient cultures and modern fields, the reader may examine these chapters in any sequence, perhaps beginning with topics that match his/her interest and expertise. One of the best tools in understanding each chapter will be to study the diagrams which provide a clear summary of what the reader really needs to know.

CHAPTER 1:

THE CONSTITUTION
OF THE UNIVERSE

THE RIG VEDA OF THE
ANCIENT VEDIC CIVILIZATION

The Vedic civilization is a term used to describe an ancient culture which appears to have existed in an area corresponding to the Northern regions of modern-day India. According to recent archeological evidence, this ancient culture is one of the oldest and largest civilizations in the world, predating the massive Indus Valley civilization in India by at least several thousand years (Frawley, 1986, 1994). In addition to its antiquity, the ancient Vedic civilization has the largest, most complete and unaltered collection of literary works, larger than the records of all other ancient cultures combined. For this reason the ancient texts (not their commentaries) of the Vedic civilization have become a valuable source for research and understanding of universal wisdom. In this section we will examine the wisdom contained in the most revered and important text of the ancient Vedic civilization, the Rig Veda. Our analysis of this ancient text is primarily derived from the work of Maharishi Mahesh Yogi, whose research over the last thirty years has uncovered a particularly clear restoration and understanding of the true wisdom recorded by the ancient Vedic people. A clear understanding of the information presented in this chapter will provide the foundation for everything that follows in this book. We will begin our discussion of this topic with a background description of the ancient Vedic civilization, then lead into a detailed understanding of the opening hymns of the Rig Veda which contain the fundamental template referred to as the Constitution of the Universe.

Description of the Vedic Literature

Much of what we know regarding the ancient Vedic civilization stems from its vast collection of written records which have survived for thousands of years. These literary works, known as the Vedic literature, comprise the most extensive, oldest and largest body of knowledge of any ancient civilization (Frawley, 1993). The Vedic literature, written in the language of Sanskrit, contains as we have mentioned, more original and unaltered teachings than all other ancient cultures combined, making it a valuable source for research. There are an estimated two to three million ancient texts in Sanskrit scattered throughout the world with individuals, families, ashrams, museums, and so on. Of these texts there are about a quarter of a million which have been catalogued and identified. From these there are only about 35,000 to 40,000 texts which have been systematically studied with translations and attempted interpretations. What this means is that there is still a huge amount of information which has not been looked at. One can only imagine how much knowledge and wisdom is yet to be rediscovered. The number of surviving texts in the Vedic literature is quite remarkable considering that India is a tropical country where written records decay quickly and the fact that India has endured many conquests.

Although the Vedic literature is often associated with the sacred scriptures of the Hindu religion, it contains much more than religious teachings (Frawley, 1992; Frawley, 1993). Contained within the Vedic literature is wisdom encompassing an entire way of life, including topics such as medicine, mathematics, military defense, architecture, music and a myriad of other areas. Throughout these ancient texts are also numerous descriptions of profound experiences of elevated states of consciousness, signifying that the Vedic civilization possessed an advanced spiritual understanding of the universe. Most of the texts of the Vedic literature are complete, unaltered and are still available in their original language of Sanskrit. Sanskrit, considered to be mother of all languages, is one of the few languages of the ancient world for which the original pronunciation has been preserved as a living tradition for thousands of years. It is also interesting to note that Sanskrit contains more words for "divine" and more precise terms for defining consciousness than any other language (Frawley, 1993).

In the Vedic literature the most important texts are considered to be a collection of four sets of books known as the four Vedas (Rig Veda, Sama Veda, Yajur Veda and Atharva Veda). Veda is a Sanskrit word meaning "knowledge" – knowledge about everything in the universe. All texts in the Vedic literature belong to, and are derived from, one of these four

Vedas. Among the four Vedas the first and foremost is the Rig Veda, translated as "Wisdom of the Verses," which is said to have been written down by the enlightened sage, Vyasa. The Rig Veda is not only the oldest text of the Vedic literature, but is also the most respected and important work as it contains all Vedic wisdom in a condensed form. It is not known how old the Rig Veda is since much of the knowledge contained in this text appears to have originated from an oral tradition of many generations.

In addition to the surviving written records, the hymns of the Rig Veda have also been preserved and passed down from generation to generation in an unbroken oral tradition. Even today, the Brahmins of India still recite the hymns of the Rig Veda with the same pronunciation and accent as their ancestors thousands of years ago. What has been lost however is the meaning and significance of the wisdom contained in the Rig Veda. As a result many western scholars have attempted to understand the hymns of the Rig Veda through intellectual commentaries on English translations. Unfortunately, this approach has been futile since most scholars have not been able to make sense of the literal meaning of the original writings of the Rig Veda and as a result these scholars have naively blamed the Vedic culture for being unintelligible and primitive. Thus, until recently, the content of the Rig Veda has remained an enigma to most individuals.

To comprehend the hymns of the Rig Veda, we must realize that these texts contain hidden universal truths and wisdom, rather than a mere collection of historical facts or poetic fantasies. The hymns of the Rig Veda are direct cognitions (by the ancient Vedic sages) of the laws of nature governing all objective and subjective aspects of the universe. These cognitions describe the sequential emergence of primal laws and patterns of our universe from an underlying unified basis. To the ancient Vedic seers, the fundamental mechanics of creation were cognized in their silent levels of awareness or consciousness. It is the wisdom from these ancient cognitions which form the hymns of the Rig Veda.

The Structure of Rig Veda is a "Self-Commentary"

Although there are numerous levels on which the Rig Veda can be interpreted, probably one of the clearest and most profound understandings is latent in the precise organization and numerical structure of the hymns.

Not only does this approach lend itself to objective analysis, but it is also relatively easy to comprehend and is free from ambiguities. During the last decade several scholars have begun to use this novel approach and have revealed an entirely new understanding of what is actually contained within the verses of the Rig Veda. One of the most prominent scholars and teachers in this field is Maharishi Mahesh Yogi (hereafter referred to as "Maharishi"). For the past three decades Maharishi, in collaboration with other Vedic scholars, has devoted his time to restoring and reviving the original meaning and understanding of the Rig Veda as well as the other aspects of the Vedic Literature.

According to the Vedic tradition, the verses of the Rig Veda are structured to comprise a sequential progression of elaboration. Each part of the Rig Veda is an elaborated commentary on the section which immediately preceded it. What this means is that the knowledge contained in the Rig Veda is developed and elaborated through a nested series of sequentially larger or expanded units. Maharishi has referred to this unique structural characteristic as the "Apaurusheya Bhashya" of Rig Veda (Dillbeck, 1989; Maharishi, 1992; Maharishi, 1994). The phrase Apaurusheya Bhashya is simply a Sanskrit expression which translates as the "uncreated commentary" or "self-commentary." Rig Veda is described as a "self-commentary" because every segment of the Rig Veda is a commentary or elaboration of the parts which preceded it. This commentary is inherent within the actual verses and is not something created externally.

One way to conceptualize this unique structural characteristic is to think of the Rig Veda (and the entire Vedic Literature) as an omniscient book. In such a book we could read the first letter and it would tell you everything about the universe. Then, if we wanted to obtain a deeper understanding, we could read the first word which would give us a little more detail about everything in creation. Likewise, the first sentence would give us a more elaborated version or commentary on the first word and first letter. In this fashion, the book would unfold in packages of complete knowledge and we would not need to read the entire book to know the whole story of creation. Similarly, this "omniscient" structure is the key to understanding the wisdom of the Rig Veda.

The organization of Rig Veda is based on sequentially larger units starting with letters which combine to form syllables. Syllables in turn combine to form lines (called "padas" in Sanskrit). Lines combine to produce verses (called "richas" in Sanskrit), which proceed to form stanzas (referred to as "suktas," pronounced "sooktas," in Sanskrit). Stanzas combine together to give what are known as mandalas or books. In the Rig Veda there are ten mandalas (equivalent to about 10,000 verses). From the Rig Veda the other three Vedas (Sama Veda, Yajur Veda and Atharva Veda) are derived, and subsequently the entire Vedic Literature. Each sequentially larger unit in the Rig Veda elaborates the information and meaning inherent in the smaller units which preceded it. Understanding this quantitative organization of the Rig Veda is a cornerstone to unraveling the wisdom of the ancient Vedic knowledge.

Underlying the phonetic structure of the Rig Veda just described, there is also a corresponding numerical structure. For example, there is a specific number of syllables that make up a line, and a particular number of lines which define a verse. Based on this numerical/phonetic structure, (in 1992) Maharishi discovered a fundamental pattern or template in the opening hymns of the Rig Veda which has become known as the Constitution of the Universe[2] (Maharishi, 1992; Maharishi, 1994). The Constitution of the Universe is a phrase used to describe a fundamental pattern that governs all order and intelligence. This numerical structure or template found in the opening hymns of the Rig Veda represents the Constitution of all constitutions from which the universe functions. One way to conceptualize the importance of this concept is to relate it to the constitution of a nation. All law and order which govern a nation are ultimately based on a set of principles laid out in its original constitution. Similarly, our universe is governed and structured based on a simple, yet fundamentally powerful, template that is mirrored throughout the cosmos. Our purpose in this book is to provide both the evidence and the implications for this unifying concept.

The First Syllable of Rig Veda Contains Complete Knowledge

Figure 1 and Tables 1 through 3 provide a detailed breakdown of the Constitution of the Universe as found in the opening hymns of the Rig

---

[2] Publicized January 1992 in the following: *The Toronto Globe and Mail* (Canada), January 1992; *The Ottawa Citizen* (Canada), January 1992; the *International Herald Tribune* (International), 8 January 1992; the *Financial Times* (Great Britain), 8 January 1992; the *Financial Times* (International), 8 January 1992; *The Times of India* (India), January 1992; *The Wall Street Journal – Europe* (Europe), 8 January 1992; *The Asian Wall Street Journal* (Asia), 10 January 1992; *The Wall Street Journal (USA)*, 6 January 1992; and *The Washington Post* (USA), 9 January 1992.

Figure 1. Locating the Structure of the Constitution of the Universe in the Rig Veda

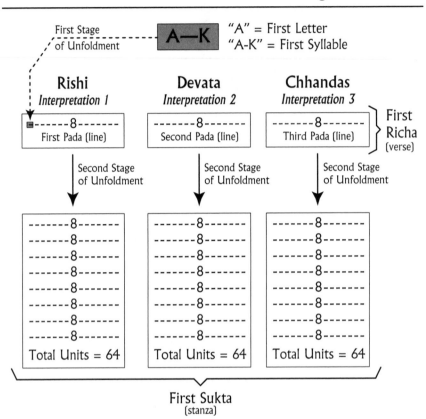

First Sukta
(stanza)

Veda. The first letter in the opening hymns of Rig Veda is "A," which is pronounced "ah" with a full opening of the mouth and throat that corresponds to its meaning as fullness. Phonetically, "A" corresponds to the least obstructed position in the physiognomy of speech. It comprises the first letter of most alphabetic systems and is thought to encompass all other sounds (Hagelin, 1989). It is said that all knowledge is contained in the letter, "A," and that each subsequent segment of Rig Veda elaborates its meaning. To the ancient Vedic sages the letter "A" represented the fullness of all possibilities and primal laws governing our universe, which reside in a unified field of consciousness accessible through direct cognition. This unified level of consciousness or awareness was believed to be the source of literally all forms and phenomena in nature.

The second letter in the Rig Veda is "K" (pronounced "ka") which represents the extreme opposite of the first letter "A." Phonetically "K" corresponds to the most obstructed and closed value of speech (Hagelin, 1989). To the Vedic sages "K" signified complete emptiness or the point value of creation.

Together "A" and "K" form the first syllable of Rig Veda "AK" (refer to Figure 1 and Table 1). This syllable "AK" represents the full range of creation from infinity ("A") to a point ("K") and everything in between. It is interesting to note that phonetically you cannot pronounce "K" without also pronouncing a suffix "A" ("ka"). According to Maharishi, the syllable "AK" is the fundamental expression of the coexistence of opposite values underlying creation

---

Table 1. The First Sukta of Rig Veda: Rishi Interpretation

The First Pada of Rig Veda *(8 Syllables)*:

Rig Veda:

अक्	नि	मीं	ळे	पु	रो	हिं	तं
AK	ni	mi	le	pu	ro	hi	tam

Prakriti:

| Ahamkara | Buddhi | Manas | Akasha | Vayu | Agni | Jala | Prithivi |

Translation:

| Ego | Intellect | Mind | Space | Air | Fire | Water | Earth |

The Elaborated Sixty-Four Aspects:

ग्र	ग्रि:	पू	वें	भि	र्ब्रं	वि	भि
ग्र	ग्रि	नां	र्	यि	मं	श्न	व्
ग्र	ग्रे	यं	यु	ज्ञ	मं	ध्व	रं
ग्र	ग्रि	हों	तां	क्	वि	क्रं	तु:
य	दु	ङ्	दां	शु	षे	तु	वं
उ	प	त्वा	ग्रे	दि	वे	दि	वे
रा	ज	न्त	म	ध्व	रा	णां	गो
स	न:ं	पि	ते	वं	सु	न	वे

---

and signifies a process by which the fullness of "A" collapses or condenses to form the point "K" (Dillbeck, 1989; Maharishi, 1992; Maharishi, 1994). This process is called "Akshara" (translated as the collapse of fullness or the collapse of all possibilities) and is said to constitute the eternal dynamics of consciousness knowing itself and contains all Vedic knowledge in seed form. To clarify this concept it may be useful to think of the relationship between a seed and a tree. Contained within a seed is all the information and intelligence needed to grow a large tree with all of its diversified structures and functions. In a similar way we could think of the first syllable of Rig Veda, "AK," as the seed of the universe containing all knowledge by which the creation is governed and controlled.

Table 2. The First Sukta of Rig Veda: Devata Interpretation

The Second Pada of Rig Veda *(8 Syllables)*:

Rig Veda:

य	ज्ञ	स्य	दे	व	मृ	त्वि	जम्
ya	jna	sya	de	va	mri	tvi	jam

Prakriti:

Ahamkara	Buddhi	Manas	Akasha	Vayu	Agni	Jala	Prithivi

Translation:

Ego	Intellect	Mind	Space	Air	Fire	Water	Earth

The Elaborated Sixty-Four Aspects:

री	ड	ग्यो	नू	तं	नै	रु	त
त्यो	ष	मे	व	टि	वे	दि	वे
वि	श्व	तः	प	रि	भू	र	सिं
स	त्य	श्वि	त्र	श्र	व	स्त	मः
अ	ग्रें	भ	इं	क	रि	ष्य	सिं
दो	षा	व	स्त	र्धि	या	व	यम्
पा	मृ	त	स्य	दी	दि	वि	म्
अ	ग्रें	सू	पा	य	नो	भं	व

Eight Primal Units Define the First Level of the Constitution of the Universe

The process of "A" condensing or collapsing to "K," as revealed by Maharishi, occurs in eight successive steps (Maharishi, 1986; Maharishi, 1994). These eight successive steps are separately expanded to form the first eight syllables of Rig Veda. The first eight syllables, which include the syllable "AK," make up the first pada or line of Rig Veda (refer to Figure 1 and Table 1). The first pada of eight syllables thus emerges from, and provides a commentary on, the first syllable "AK." In Rig Veda the eight

---

Table 3. The First Sukta of Rig Veda: Chhandas Interpretation

The Third Pada of Rig Veda *(8 Syllables)*:

Rig Veda:

य	ज्ञ	स्यं	दे	व	मु	त्वि	जर्म्
ho	ta	ram	ra	tna	dha	ta	mam

Prakriti:

| Ahamkara | Buddhi | Manas | Akasha | Vayu | Agni | Jala | Prithivi |

Translation:

| Ego | Intellect | Mind | Space | Air | Fire | Water | Earth |

The Elaborated Sixty-Four Aspects:

री	ड	ग्यो	नू	तं	नै	रु	त
त्यो	ष	मे	व	दि	वे	दि	वे
वि	श्व	तंः	प	रि	भू	र	सिं
स	त्य	श्रि	त्र	श्र	व	स्त	मः
अ	ग्रे	भ	द्रं	क	रि	ष्य	सिं
दो	षा	व	स्त	र्धि	या	व	यम्
पा	मु	त	स्य	दी	दि	वि	म्
अ	ग्रे	सू	पा	य	नो	भ	व

syllables forming the first pada are said to represent the eight primal tendencies of Nature and correspond to what are known as the eight prakritis. The nearest English word describing prakriti is "nature" or "fundamental quality of intelligence" (Maharishi, 1967, Chapter 3, verse 5). At the top of Table 1 we have presented the eight syllables of the first pada along with the corresponding eight prakritis. Below each prakriti is the English translation according to Maharishi. It should be emphasized however that because English words are typically not able to capture the full meaning of a Sanskrit word, each of these English translations serves only to give a general feeling or flavor of the qualities embodied by that prakriti.

The first three prakritis are referred to as the three subjective aspects of life. Ahamkara (translated as ego) is the first prakriti which is defined as self-concept, individual consciousness, self-identity, soul, purity, subtlety and sense of "I-ness." The second prakriti is Buddhi (translated as intellect). Qualities characterizing the intellect are logic, discrimination, evaluation, reason, analysis, precision, sharpness and decision making. The third prakriti is Manas (translated as mind). Characteristics of the mind include creativity, thinking, reflecting, contemplating, recollecting and planning. Other aspects also included in the mind are memory, mental activity, attention, joy and sorrow. Collectively these first three prakritis are responsible for all subjective aspects in our universe.

The remaining prakritis (4 through 8) comprise what are known as the five "tanmantras" or five subtle elements: Akasha (space), Vayu (air), Agni (fire), Jala (water) and Prithivi (earth). As alluded to earlier, the English translations for each of these prakritis serve only to give an idea or flavor of what that prakriti really represents. For example, the prakriti Agni (fire) would embody qualities such as heat, brightness, digestion, lightness, destruction and sharpness, rather than simply the image of burning wood. Similarly, the qualities of the other subtle elements are embodied in the characteristics associated with space, air, water and earth, respectively. Together these five subtle elements are responsible for all objective and material aspects of our universe. Interestingly, it appears that the understanding and knowledge of these five elements was ubiquitous throughout the ancient world. In fact, the second half of this book is devoted to describing a developmental process by which these five basic elements begin to construct the material diversity of creation.

So far we have described the first stage of unfoldment of the Constitution of the Universe (refer to Figure 1). From the first syllable, "AK," representing unity or all knowledge in seed form, arise the first eight syllables or eight prakritis of Rig Veda which correspond to the eight-fold divided nature of consciousness or eight primal divisions of creation. These eight basic units arising from one state of unity govern the construction and diversification of any system within our universe.

The Constitution of the Universe Contains Three Angles of Interpretation

The next stage of structure in the Rig Veda is the construction of the first richa (verse) from the first pada of eight syllables (eight prakritis). In this construction the eight prakritis appear sequentially three times to create a total of three padas (eight syllables each) which together make up the first verse of Rig Veda (twenty-four syllables in all). The eight syllables in each pada provide a different interpretation and description of the eight prakritis. We are not creating any new entities, but rather describing in three different ways or flavors the eight prakritis we already have (creating a total of twenty-four syllables which define the first verse). In this way the twenty-four syllables of the first verse of Rig Veda provide a further commentary on our original eight syllables, as follows.

The original pada of eight syllables describes the eight prakritis with respect to what is called the rishi interpretation. Rishi is a Sanskrit word which means "knower," "observer," "seer" or "silent witness." Qualities often associated with the rishi interpretation include abstract, inherent nature, subjective and intelligence. The second pada of eight syllables characterizes the eight prakritis in relation to what is known as the devata interpretation. The Sanskrit word devata is defined as "process of knowing" and "process of observing." It is often denoted by qualities such as dynamism, activity, transition, change, interaction and movement. The eight syllables in the third pada describe the eight prakritis with respect to chhandas. Chhandas is translated as "known" or "observed." It embodies characteristics such as material, concrete, physical, expressed and worldly.

Collectively, rishi (knower), devata (process of knowing) and chhandas (known) are said to be the three aspects of knowledge or the primal interpretations necessary to gain full understanding of an entity. According to Maharishi, these three aspects of knowledge are derived from the very nature of consciousness itself (Maharishi, 1986; Maharishi, 1994).

Consciousness, simply by virtue of its own existence, is conscious or aware. This state where consciousness is conscious is called the rishi (the knower) aspect. The question then arises, "What is consciousness aware of?" If consciousness is the ultimate reality of all that there is in the universe, then consciousness can only be conscious of itself. When consciousness is conscious of itself, this is referred to as chhandas (the known aspect). In order for consciousness (the knower) to know itself (the known) there has to be a process of knowing (devata) to connect the rishi aspect and chhandas aspect. These three intellectual concepts or flavors of intelligence are expressed throughout the diversified layers of creation, including the primal level of the eight prakritis, thereby forming the three padas of the first verse of Rig Veda. In the Vedic Literature these three aspects of knowledge collectively are called the samhita (pronounced sang-heeta) value or the complete description and understanding of an object. Samhita, therefore, is the togetherness of rishi, devata and chhandas.

For simplicity we have provided a detailed breakdown of the construction of the first verse of Rig Veda in Figure 1 and Tables 1 through 3. In actuality however, the way this verse appears in the texts of Rig Veda is as follows:

Agnim[3] ile purohitam	Pada 1 (8 syllables) – Rishi interpretation of 8 Prakritis
Yagyasya devam ritvijam	Pada 2 (8 syllables) – Devata interpretation of 8 Prakritis
Hotaram ratna dhatamam	Pada 3 (8 syllables) – Chhandas interpretation of 8 Prakritis
	Verse 1 (24 syllables) of Rig Veda

From this first verse we can already see how naive scholars could easily find the Rig Veda unintelligible if they were to base their analysis solely on the literal meaning of an English translation. As we have described previously, the real knowledge of the Rig Veda is concealed in its numerical structure which, when really understood, sheds an entirely new light on the literary words and syllables.

---

[3] According to rules of Sanskrit grammar, the syllable "AK" becomes "Ag" because of the letter "n" that follows it in the word *Agnim* (Dillbeck, 1989).

Sixty-Four Defines the Second Level of the Constitution of the Universe

The final level of structure or elaboration in the Constitution of the Universe as located in the Rig Veda is the construction of the first sukta (stanza) from the first richa (verse). In this construction the first verse of twenty-four syllables is elaborated or expanded eight-fold to create an additional eight verses. These eight new verses plus the original verse together make up what is called the first sukta or stanza of Rig Veda (refer to Figure 1). Each of these eight new verses, as in the first verse, contain twenty-four syllables divided equally into three padas. Thus, collectively the eight new verses contain twenty-four (8 × 3) padas of eight syllables each or a total of 192 syllables.[4]

As in the first verse, each of the three padas in the eight new verses of the first sukta, respectively present a description or interpretation with respect to rishi, devata and chhandas. As a result, the 192 syllables making up the eight elaborated verses of the first sukta are essentially divided into three groups of sixty-four syllables (64 × 3 = 192 syllables). The first group of sixty-four syllables represents an eight-fold elaboration on the first pada of Rig Veda or the rishi (knower) interpretation of the eight prakritis (refer to Figure 1 and Table 1). Likewise, the second group of sixty-four syllables constitutes an eight-fold elaboration on the second pada of Rig Veda or the devata (process of knowing) interpretation of the eight prakritis (refer to Figure 1 and Table 2). Similarly, the third group of sixty-four syllables delineates an eight-fold elaboration on the third pada of Rig Veda or the chhandas (known) interpretation of the eight prakritis (refer to Figure 1 and Table 3).

For the purpose of this book we have omitted a detailed analysis of the three groups of sixty-four syllables in the first sukta of Rig Veda. Rather, we have simply presented these 192 syllables (without English translations) in Tables 1 through 3. Our primary interest at this stage is simply the numerical structure of the opening hymns of the Rig Veda. Further analysis of literal meaning of each of the 192 syllables is possible, but is

---

[4] According to Maharishi these twenty-four padas of eight syllables represent an elaborated eight-fold structure of the twenty-four gaps between the twenty-four syllables of the first verse.

tedious, and is not necessary in order to comprehend the significance of
the Constitution of the Universe.

A Recapitulation of the Constitution of the Universe

At this stage we have completed the description of the Constitution of the
Universe as located by Maharishi in the first sukta of Rig Veda. Additional
elaboration and other structural patterns are present in later parts of the
Rig Veda, but serve only to comment in greater detail on the fundamental
patterns already inherent in the opening hymns. For example, the 192
syllables of the eight elaborated verses of the first sukta are further elaborated to create the 192 suktas constituting the first mandala or book of Rig
Veda. The first mandala of Rig Veda is then further elaborated into the ten
mandalas of the complete Rig Veda. These ten mandalas of Rig Veda in
turn provide the foundation for an even more extended commentary
inherent in the other three Vedas (Sama Veda, Yajur Veda and Atharva
Veda) and subsequently the entire Vedic Literature. Contained in the more
elaborated texts of the Vedic Literature is detailed wisdom encompassing
all disciplines as well as an entire way of life. However, the significance of
the first sukta of Rig Veda is that it contains all the wisdom and knowledge of the Vedic Literature in seed form. It is a fundamental constitution,
hence the "Constitution of the Universe," containing the primal patterns
and mechanism which create and govern all the details and diversity of
our universe.

To ensure a thorough comprehension of the numerical nature of the
Constitution of the Universe we suggest re-reading this section and
studying Figure 1, which will become the reference diagram throughout
the first part of this book. It will also be helpful to understand the literal
meaning of the structure of the Constitution of the Universe as expressed
in the Rig Veda (e.g., the eight prakritis and the three interpretations of
rishi, devata and chhandas).

To summarize, the primary structure of the Constitution of the Universe
consists of two stages. The first stage consists of a fundamental set of eight
units (the eight prakritis) which emerge from a state of unity (the first
syllable, "AK"). In the second stage, this fundamental set of eight units is
elaborated eight times to create sixty-four units (the sixty-four syllables in

the first sukta). Secondarily, at each of these two stages there is a triple-flavor interpretation with respect to the three fundamental aspects of knowledge: rishi (knower), devata (process of knowing) and chhandas (known). Although this numeric/phonetic pattern of numbers may seem trivial and simple, it will soon become evident that this structure is the fundamental template which is mirrored throughout the universe.

Validation of the Constitution of the Universe

Prior to Maharishi's description of the Constitution of the Universe located in the opening hymns of the Rig Veda, most attempts at understanding this aspect of the Vedic literature were futile. As stated earlier, the primary reason why the Rig Veda has remained an enigma for so long is that most scholars have been casual readers and have tried to interpret this knowledge from the level of superficial translations. It is not enough to have the Vedic literature if we cannot comprehend it. Nor is it sufficient to translate the texts into English if we do not know the spiritual background in which they were written (Frawley, 1986; Frawley, 1994). What we need to realize is that, even though on the surface the hymns of the Rig Veda may appear like poetry or prayers, there is a well-defined numerical structure and deeper philosophical meaning hidden in these texts. Interestingly, other individuals are also beginning to find significant patterns in the hymns of the Rig Veda. For example, in the last several years Subhash Kak (Kak, 1993a, 1993b and 1994; Kak and Frawley, 1992) discovered that very accurate and precise astronomical numbers such as planetary periods and distances between planets are embedded within the structure of the Rig Veda. Also, it is believed that every letter in the Rig Veda corresponds to a particular code number. If we decode certain hymns using a particular alpha-numeric coding system, it is possible to derive numbers such as pi up to thirty-two digits and even beyond. We are in the earlier stages of discovering the deeper wisdom and knowledge hidden in the Rig Veda and other aspects of the Vedic Literature. The doors of research are wide open and much information is still to be rediscovered.

Shortly after Maharishi presented his description of the Constitution of the Universe located in the opening hymns of the Rig Veda, he proceeded to substantiate the validity of this template by collaborating with several physicists to reveal this exact same numerical pattern underlying the mathematical equations of the unified quantum field theories in modern physics (refer to Chapter 2, Section I). This intriguing similarity between the ancient Vedic wisdom of the primal mechanics of creation and the currently accepted theories about the origin of the universe by modern physics hinted at the significance of the Constitution of the Universe. Since this initial work with the Constitution of the Universe, little research has been done and the real significance of this template has for the most part gone unnoticed. However, as a result of our research during the past two years, we feel that we have definitive evidence to demonstrate that the numerical structure described in the Constitution of the Universe is not only found everywhere, but also describes the fundamental mechanisms everywhere. The following chapters of the first part of this book provide nineteen illustrations of this pattern found across the academic disciplines and throughout the ancient civilizations.

CHAPTER 2:

PHYSICS AND THE CONSTITUTION OF THE UNIVERSE

SECTION I

QUANTUM MECHANICS: THE $E_8 \times E_8$ HETEROTIC SUPERSTRING THEORY

In recent years the primary challenge in theoretical physics has been to construct mathematical formulations that attempt to explain or accommodate the rich diversity of forces and particles observed in our universe as arising from a single unified source. These mathematical equations, known as unified field theories, provide one of the most promising frameworks in the history of Western science that may ultimately fulfill Einstein's quest of finding a completely unified understanding of all entities in creation. One of the most successful and promising unified field theories to emerge from these efforts is called the $E_8 \times E_8$ heterotic superstring theory. In this section we provide a conceptual understanding of superstring theory and present evidence that the mathematical structure of this unified field theory includes the basic numbers eight, sixty-four and three of the Constitution of the Universe. This correspondence was originally discovered by Maharishi in collaboration with physicists Dr. John Hagelin and Dr. Robin Ticciati of Maharishi International University (Hagelin, 1992; Kelley, 1993; Maharishi, 1992; Maharishi, 1994).

Emergence of Quantum Mechanics from Newtonian Physics

Prior to the 20th century, physicists believed that the laws of classical mechanics explained everything about the material universe. In classical mechanics (also known as Newtonian physics) forces and objects are governed by rational laws which are deduced from sensory observation of the everyday world. Accordingly, nature was assumed to run its own course in space and time conforming to discrete laws that were easy to understand and simple to picture. The universe was viewed in terms of solid bodies, straight-line motion and fixed constants that governed the destiny of all events. For instance, the structure of an atom was thought to resemble that of a solar system, in which the electrons revolved like planets around a central nucleus. The hallmark of classical mechanics is that accurate predictions and explanations can be given for macroscopic objects moving at speeds well below the velocity of light.

Around the turn of the century, however, physicists began to realize that the classical laws of Newtonian physics were inadequate to explain the events observed at the molecular, atomic and subatomic levels of creation. Between 1870 and 1910, a variety of phenomena were observed at these deeper realms of nature that defied the commonsense notions and laws that applied to the world of large-scale objects. Out of these observations emerged a new branch of physics known as quantum mechanics, which was pioneered by Planck, Einstein, Bohr, Heisenberg and other leading physicists. Quantum mechanics provided a new language of nature to explain the laws governing the phenomena of the subatomic levels which are beyond the limit of sensory experience.

Fundamentally, quantum mechanics differs from Newtonian physics in that energy is no longer considered to be a continuous variable, but rather is constrained to having discrete states or quanta. Under certain conditions, this discreteness gives rise to the particulate appearance of the material universe which we interpret as the elementary particles of quantum physics. Another significant advance in quantum mechanics is called quantum field theory. This is fundamentally a theory of fields that presents a simple and profound view, in which the unrelated concepts of particles and forces are unified and explained in terms of an underlying quantum field. Quantum field theory has been successful in explaining the behavior of particles in relation to field interactions, as well as in predicting the existence of new types of particles.

During the first half of this century quantum physics witnessed a proliferation in the number of elementary particles. Prior to that time the atom was thought to consist of three types of particles (electrons, protons and neutrons). However, as physics advanced in technology, hundreds of additional subatomic or elementary particles were discovered, leading to the search for systematic relationships among the various components. Over the past few decades this quest has led quantum physicists to progressively more unified levels of nature's functioning. In search of more unified principles, physicists have formulated mathematical theories to combine the forces of nature with the fundamental particles. The ultimate goal of these efforts is to find a unified field theory which is consistent within itself and which is capable of explaining literally everything.

Historical Development of Unified Field Theories

The first step toward a unified field of nature was revealed by Michael Faraday in the 1830s on the intimate connection between the electric force and the magnetic force. Not only did Faraday believe that there was a unity between electricity and magnetism, but he also envisioned that there must be a unified field at the basis of the universe. Building on Faraday's work, James Clerk Maxwell eventually succeeded in formulating a set of equations that unified the electric and magnetic forces into one unified force called electromagnetism. Electromagnetism is seen as one of the four fundamental forces of nature and is responsible for atomic structure, chemical bonding, electronic technology and the spectrum of electromagnetic radiation (e.g., x-rays, light, radio waves, microwaves, etc).

Further progress towards a unified understanding of nature began in 1967 with the successful unification of the weak and electromagnetic forces into what is called the electroweak force. (The weak force underlies the process of energy production in stars, radioactive decay and neutrino interaction.) This unification of the weak force and electromagnetic force was discovered through the brilliant work of Glashow, Salam and Weiberg (who were jointly awarded the Nobel Prize in 1979). According to this electroweak theory, the electromagnetic force and weak force assume identical physical properties. These two fields become fundamentally indistinguishable from one another, meaning that they are one unified field. As a result of this research, the number of forces needed to explain all phenomena in the universe was reduced to three: the electroweak force, the strong force and gravity.

Although the electroweak field obviously does not exist in our present day universe, experimental evidence for the existence of such a field has been obtained from an advanced technology known as particle acceleration. In particle accelerators, particles are accelerated to very high speeds and then collided with one another. The fragments of such collisions provide evidence for the existence of subatomic particles and more unified levels. In Figure 2 the electroweak force corresponds to a distance scale of about 10^{-16} cm. It is also theorized that the electroweak force played a vital role in the development of our universe during the first one-billionth of a second following the postulated Big Bang. During this very short period the physical environment of our universe was suitable (e.g., high temperature and compactification) for the natural existence of the electroweak force.

A deeper and more profound level of unification of the fundamental forces and particles occurs in the context of what is known as grand unification. The term grand unification denotes a unified field theory which attempts to unify the weak, electromagnetic and strong interactions. (The strong force is a field that acts over very short distances to hold together the structure of the atomic nucleus.) At this level of unification all the elementary particles and forces (except gravity) are united and fundamentally indistinguishable from one another. Although a number of theoretical models have been proposed for grand unification, experimental verification has been difficult to obtain due to the extremely small time and distance scales (in Figure 2 this corresponds to 10^{-29} cm) and correspondingly high energy levels. This is one of the major problems confronting physics today. For example, using current technology to probe 10^{-29} cm would require a particle accelerator about the circumference of our galaxy.

The ultimate level of unification of the fundamental particles and forces occurs in the context of super unification. The term super unification signifies a completely unified field theory that unites the four fundamental forces of nature into a single field that represents the totality of natural law and explains the rich emergence of diverse particles and forces. Mathematical formulations of completely unified field theories provide an objective understanding of the elementary particles and forces as stable modes or vibrational states of the field. However, because of the extremely small space-time distance scale of super unification, there is often little experimental evidence available to support these theories. As a result, much of the research in this area of physics has been purely theoretical

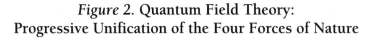

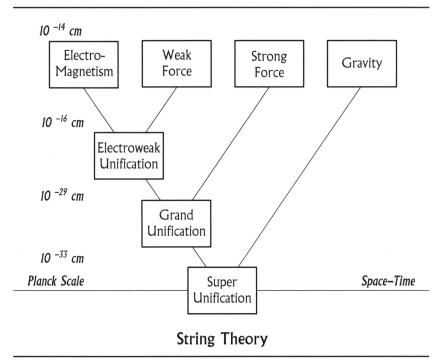

Figure 2. Quantum Field Theory:
Progressive Unification of the Four Forces of Nature

and based on speculation and hunches. What this means is that theoretical physicists have had to resort to constructing mathematical theories and then calculating whether such formulations can predict and explain the observed phenomena of our universe.

According to quantum physics, super unification occurs at the most basic level of creation, known as the Planck Scale. The Planck Scale is intrinsic to nature and describes the smallest possible dimensions. At the level of the Planck Scale everything is united into a single field which means that our classical notions of space and time become meaningless. It is impossible to measure anything at that level because nothing exists in isolation and everything is unified. Based on experimental evidence, the physical dimension of the Planck Scale is equal to about 10^{-33} cm (refer to Figure 2) and has an energy equivalent to 1.2×10^{19} GeV/c^2. Such conditions are theoretically believed to have existed in our universe until about 10^{-43} seconds after the Big Bang.

Progress towards constructing completely unified field theories began in 1974 when a profound mathematical symmetry known as supersymmetry was introduced to quantum physics, which was capable of unifying force fields and matter fields. Since that time the application of supersymmetry has led to the development of mathematical formulations that attempt to account for all the fundamental particles and forces observed in our universe. One of the first comprehensive unified field theories to emerge from these formulations was the $N = 8$ supergravity theory containing a maximum of eight supersymmetries. However, certain basic problems and predictions of the supergravity theories have been found to be incompatible with experimental observations. In recent years those earlier theories have become replaced by what are called superstring theories that may ultimately provide the answer to finding a completely unified field theory.

Superstring Theory is the Most Promising Unified Field Theory

In superstring theory, the structure and internal dynamics of a completely unified field can be described in three general ways: Hilbert Space, Operators on Hilbert Space and particular states in Hilbert Space. The term Hilbert Space refers to an abstract field or an infinite dimensional space, where all creation takes place, and may be viewed as the arena where all quantum mechanical functions live and exist. It is an abstract vector space of all possibilities that can be expanded and interpreted in any other state. Operators on Hilbert Space (e.g., energy, angular momentum, etc.) function as dynamic generators of transformation in Hilbert Space. They actively transform one quantum-mechanical state into another and are dynamic generators of all change governing our universe. Without these Operators, and the transformation they generate, Hilbert Space would be completely inert. Living within Hilbert Space are particular states or individual points that represent isolated possibilities. Each state corresponds to an actual mode of the physical system. Together these three aspects of the unified field provide a profound and simple view of the quantum levels of life residing below the molecular, atomic and subatomic levels of creation.

The most successful and promising string theory to emerge in recent years is the $E_8 \times E_8$ heterotic superstring theory. Conceptually this is a quantum theory whose fundamental excitations are analogous to closed elastic strings, akin to a rubber band. In a closed elastic string (which is a one-dimensional extended object) the two ends are tied together to form a continuous unbroken loop. Below the Planck Scale all elementary particles and forces correspond to massless vibrational states of the string or string fundamentals. At higher energy modes or "harmonics" the string corresponds to heavier particles with masses. Essentially this depicts our universe as being made up of tiny strings. The strings postulated by this theory are about 10^{-33} cm long (equivalent to the Planck length) and the energy required to excite a string is in the order of the Planck energy. Hence, at ordinary time and distance scales of the macroscopic world the string is not excited and appears to behave as a point particle.

A string has a rich spectrum of internal modes of vibration that correspond directly with the elementary particles and forces. Prior to the development of string theory, quantum field formulations were based on the concept of the classical point with zero dimensions. A classical particle is a point-like object that has no internal structure and is confined to move only through space. Thus, in order to explain the emergence of diversity, physicists had to assign arbitrary qualities to this classical point and then see whether it matched the observed particles and forces of nature. However, in superstring theory, a string does not need assigned qualities because it has qualities of its own. For example, a string is endowed with the freedom to vibrate in a number of transverse dimensions perpendicular to the string. One of the strengths of superstring theory is that these inherent vibrations match up very well with the actual particles and forces we observe in nature.

One of the most profound ways to understand string theory is to imagine the one-dimensional string moving in space-time: as it moves, it naturally sweeps out a two-dimensional surface called the world-sheet. Since this two-dimensional surface has the unique property of infinite symmetry, there are an infinite number of ways to transform the coordinates without inducing any change. All the quantum fields including space and time reside on the world-sheet . In other words, space and time may be viewed simply as fields residing in the world-sheet, and these fields later become interpreted by the human psychology as the world around us. This concept is called the "world-sheet view of the string."

There are strong constraints for the types of fields which can live on the world-sheet. For mathematical consistency, it turns out that there are essentially only two types: the bosonic string and the fermionic string. Bosonic strings have traditionally been associated with "forces" and tend to occupy that same quantum state. By occupying the same quantum state, bosonic strings are responsible for coherent collective behavior (e.g., super conductivity, superfluidity and laser light). Most familiar things, such as the four fundamental forces of nature, are derived from bosonic strings. In contrast, fermionic strings have been associated with "particles" and are forbidden to occupy the same quantum state according to the Pauli exclusion principle. As a result, fermionic strings do not exhibit collective coherence and are thus responsible for diversification. Together bosonic and fermionic strings form the two fundamental classifications of matter in quantum physics.

A Conceptual Understanding of Superstring Theory

According to string theory, the most compact formulation of the world-sheet, which occurs at the level of super unification, has two modes of vibration: right-movers and left-movers (refer to Figure 3). For mathematical reasons and quantum mechanical consistency, there are a total of twenty six bosonic fields which make up the right-movers. Conceptually this corresponds to the string moving in twenty-six dimensions. In the left-movers there are a total of ten superfields which describe the string living in a ten dimensional superspace. A superfield is a pair of bosonic and fermionic fields which are unified through the mathematical concept known as supersymmetry. There is no way for us to imagine the heterotic string moving in space and time because half of it is moving in twenty-six dimensions and the other half in ten dimensions. The best way to think of the string is to know that all these different dimensions are functions that are derived from the world-sheet.

The twenty-six bosonic right-movers are considered to provide an "embedded space" where the string lives. They provide the space-time arena in which the world-sheet exists. In contrast, the ten fermionic left-movers describe the characteristics of the heterotic string and the qualities of the universe as it exists at this level of nature's dynamics. For this reason, fermionic fields are considered to be very important and play a major role in constructing our 3 + 1 dimensional structure of space-time. Mathematically there are exactly eight fermionic physical degrees of

Figure 3. Summary of the Number of World-Sheet Fermions in the Mathematical Structure of the Heterotic Superstring

8 Fermions in Maximal Dimension

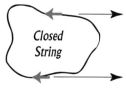

Closed String

Right Movers: 26 dimensional space
(1 dimension of ordinary space = 1 boson)

Left Movers: 10 dimensional superspace
(1 dimensional superspace = 1 boson + 1 fermion)

- A string in Nd space has N-2 physical, transverse degrees of freedom

- Total of 10 - 2 = 8 fermionic degrees of freedom, all left movers

64 Fermions in 4d Space-Time

- The formulation of the string in 4d requires the elimination of extra dimensions: 26 - 4 = 22 extra right-moving dimensions and 10 - 4 = 6 extra left-moving dimensions

- Mathematically each boson (which gets interpreted as an extra dimension) can be traded for 2 fermions

- 22 extra left-moving bosons and 6 extra right-moving bosons = 2 x (22 + 6) = 56 fermions

- Total: 56 + original 8 = 64 fermionic degrees of freedom

Each Fermionic Field Represents 3 Values

- Abstract Hilbert Space   *Rishi Interpretation*

- Operator on Hilbert Space   *Devata Interpretation*

- Particular State in Hilbert Space   *Chhandas Interpretation*

Adapted from Kelley, S. (1993). A Symposium on the Constitution of the Universe: World-Sheet Fermions and the Syllables of the Ved. *MIU Video Magazine, Maharishi International University,* Volume 6, Tape 3.

freedom intrinsic to the string in maximal dimension. To calculate this number, we simply add up the number of fermionic fields in the world-sheet and subtract two. Two is subtracted because for any n-dimensional space there are only n - 2 ways in which the string can vibrate (the two dimensions which are subtracted include time which does not vibrate and the stretching vibration which does not exist due to the internal symmetry of the world-sheet). Since the only fermionic fields that exist on the world-sheet are the ten left-movers, there are 10 - 2 = 8 fermionic degrees of freedom. These eight degrees of freedom are actually different fields of the string which are directly relevant to explaining the physical world.

To explain our present day universe, which has a 3 + 1 dimensional structure of space-time, we need to "compactify" or "fermionize" the world-sheets in maximal dimension. To do this, we obviously need to get rid of, or hide, the "extra" dimensions. The most elegant and accepted mechanism for doing this is called the free-fermionic formulation of the string in four dimensions. In its simplest version, this procedure involves "trading" or "switching" different fields. According to superstring theory, every bosonic field on the world-sheet is equal to two fermionic fields. Hence, in order to account for our bosonic 3 + 1 dimensional structure of space-time, we can simply trade each of the extra bosonic fields for fermionic fields. These fermions then become interpreted as internal degrees of freedom. When we do this, we end up with 22 extra right-moving bosonic strings (26 - 4 = 22) and 6 extra left-moving superfields (10 - 4 = 6). Thus, the total number of extra bosonic strings will be 22 plus 6 which equals 28. Trading each of these 28 bosonic strings for fermionic fields gives a total of 56 fermionic fields. Adding these 56 to the original 8 fermionic degrees of freedom gives a total of sixty-four world-sheet fermionic degrees of freedom (refer to Figure 3).

These sixty-four fermionic degrees of freedom that exist in our 3 + 1 dimensional structure of space-time are the fundamental building blocks that interact and combine to create all the elementary particles and forces. In building string theories, what physicists do is to choose numbers or values for each of the sixty-four fermionic fields and then observe how the calculations match the experimental observations of our universe. So far, string theory appears to be very successful in accounting for all particles and forces as well as in solving some of the major problems of earlier unified field theories.

The Formulations of Superstring Theory are Identical to the Constitution of the Universe

We have now presented a basic conceptual understanding of superstring theory and clarified some of the fundamental numbers used in these mathematical formulations. Using this knowledge, we will proceed to locate in the structure of the $E_8 \times E_8$ heterotic string theory various correspondences with the wisdom recorded in the ancient Vedic literature. Our first correlation is based on one of the branches of the Vedic literature known as the Shrimad Devi Bhagavatam. In this aspect of the Vedic literature one particular verse describes a universal field of consciousness underlying all creation, called atman (meaning "self"). According to this verse, atman has the characteristic of sutratma (literally, "string-self"):

> *mama chaiva shariram vai sutram ity abdhidiyate* (III.7.41)

> "My body is called a string."
> *or*
> "This my body having the nature of a string."

The sutratma or string has the shape of circle or mandala which matches intriguingly well with the concept of the closed heterotic superstring developed by modern quantum field theorists. This parallel and translation were reported by Dr. John Hagelin of Maharishi International University (Hagelin, 1989).

In addition to this qualitative correspondence, there is also an exact numerical correspondence between the mathematical formulations of superstring theory and the numbers eight, sixty-four and three of the Constitution of the Universe as described in the Rig Veda (Hagelin, 1992; Kelley, 1993; Maharishi, 1992; Maharishi, 1994). Our first connection relates to the first level of unfoldment of the Constitution of the Universe, from the first syllable, "A-K," into the eight syllables or prakritis of the first pada (refer to Figure 1). Likewise, in the $E_8 \times E_8$ heterotic string theory there are exactly eight world-sheet fermionic degrees of freedom in maximum dimensional space-time. These eight fundamental fields or physical modes describe the character of the heterotic string and the peculiarity of the universe that exist at the level of super unification. For these reasons, the eight fermionic degrees of freedom perfectly match the eight prakritis of the Constitution of the Universe.

From these eight basic syllables of the first pada of Rig Veda, the next stage of expansion in the Constitution of the Universe is the interpretation of each of the eight syllables with respect to rishi (knower), devata (process of knowing) and chhandas (known). This produces a total of twenty-four units. Likewise each of the eight world-sheet fermionic degrees of freedom are interpreted with respect to the three aspects of the unified field which are consistent with the quantum-mechanical structure of string theory. As described earlier, the first way to describe a superstring field is called abstract Hilbert Space. Hilbert Space is an infinite dimensional space where all quantum mechanical functions live and exist. It is an abstract vector space of all possibilities, which corresponds well with the subtle and abstract nature of rishi. In the second interpretation, each string field is viewed as an Operation on Hilbert Space. These Operators function as dynamic generators of transformation and change in our universe. For these reasons, it makes sense to match the Operators in Hilbert Space with the dynamic and active nature of devata. The third interpretation perceives each string field as a particular vibrational mode or state in Hilbert Space. A particular state of Hilbert Space represents an isolated possibility within the quantum mechanical field of all possibilities. Each state corresponds to an actual mode of the physical system. This matches the material nature of chhandas (Hagelin, 1987; Hagelin, 1989). Together these three interpretations, when viewed with respect to the eight fermionic degrees of freedom, produce a twenty-four-fold structure that correlate precisely with the twenty-four syllables of the first richa (verse) of Rig Veda.

In the next stage of elaboration of the Constitution of the Universe the eight syllables of the first pada of Rig Veda expand into sixty-four units. Likewise in the free-fermionic formulation of the string in four dimensions, all bosonic fields of the string in maximum dimensional space-time are fermionized in order to construct our 3 + 1 dimensional structure of space-time. This process yields exactly sixty-four fermionic degrees of freedom. These sixty-four degrees of freedom are the primal building blocks of the entire physical universe which, through various combinations, create all the particles and forces in our four-dimensional structure of classical space-time geometry.

The final correspondence with the Constitution of the Universe is the interpretation of each syllable of the sixty-four-fold structure of Rig Veda with respect to rishi, devata and chhandas. This creates a total of 192 units. Similarly, when the sixty-four fermionic string fields are interpreted with respect to Hilbert Space, Operators on Hilbert Space and particular states in Hilbert Space, this gives 192 fundamental expressions.

In conclusion, we have presented a conceptual understanding of the $E_8 \times E_8$ heterotic string theory which is currently the most promising framework for a completely unified field theory. Although there is still much work to be done in quantum field theory, we can already see that the basic numerical structure of superstring formulations precisely matches the pattern displayed in the Constitution of the Universe. Specifically, the eight prakritis correspond to the eight world-sheet fermions in maximal dimension, and the expanded sixty-four syllables of Rig Veda match the more expressed sixty-four fermionic degrees of freedom in our 3 + 1 dimensional structure of space-time. Further, the qualities of rishi, devata and chhandas are identical to the three aspects of a quantum field: abstract Hilbert Space, Operators on Hilbert Space and a particular state of Hilbert Space. This exact match between the mathematical structure of string theory and the Constitution of the Universe suggests that these formulations reflect a universal pattern that may ultimately provide a completely unified understanding of the fundamental particles and forces of nature.

SECTION II

THE ELECTROMAGNETIC SPECTRUM

Light from our sun, radio transmission, x-rays used by dentists, brain waves and the energy used in microwave ovens to cook food are all examples of what physicists have termed electromagnetic radiation. Since the beginning of time, our universe has been a sea of these electromagnetic waves, and through the advent of science the knowledge of this phenomenon has formed the backbone of our modern technology. In this section we will discuss the basic principles of these electromagnetic waves, which collectively are known as the electromagnetic spectrum. Based on the full range of the electromagnetic spectrum, we propose a numerical structure or categorization matching the pattern displayed in the Constitution of the Universe.

Description of Electromagnetism

The existence of electromagnetic waves was first deduced by James Clerk Maxwell in the 1860s with his work on linking together electricity and magnetism. From his mathematical equations Maxwell was able to unify all electric and magnetic phenomena into one field called electromagnetism (one of the four forces of nature). According to his equations, a magnetic field would be produced in empty space if there were a changing electric field. Building on the idea of symmetry, he demonstrated that the reverse was also true, that is, a changing magnetic field will produce an electric field. Hence, if we have a changing electric field producing a changing magnetic field, this changing magnetic field will then in turn produce a changing electric field, and so on. The interaction of these two fields will actually propagate through space and is responsible for the production of electromagnetic waves.

In the current conditions of our universe, all atoms in every substance are constantly moving and being jostled. As a result, the electrically charged electrons in these atoms are always moving, automatically creating waves of electromagnetic radiation. These electromagnetic waves are thus constantly emitted by all objects. One of the functions of electromagnetic radiation is to carry energy through space, even though mass itself does not travel the same distances. For example, the radiant energy from our sun provides a source of light and heat energy for our planet.

Each electromagnetic wave has three primary characteristics: wavelength, frequency and energy. The wavelength (symbolized by λ) is the distance between two consecutive crests or between any two identical points on the propagating wave (also referred to as a cycle). Frequency (represented by v) describes the number of full wavelengths or crests that pass a given point per unit time and is usually measured in hertz, or cycles per second. Electromagnetic waves exhibiting a high frequency have a short wavelength, and electromagnetic waves with a low frequency have a long wavelength. This inverse relationship between frequency and wavelength is expressed in the following formula (where c is equal to the speed of light):

$$\lambda \, v = c$$

This means that all electromagnetic waves travel at the same velocity, the speed of light, which is equal to 3.0×10^8 meters/second. Historically this was a major advancement because it solved the mystery of the nature of light – visible light was simply an electromagnetic wave.

In order to describe the energy of an electromagnetic wave, we need to introduce a new concept. In 1901 the German physicist Max Planck discovered that energy is "quantized" at the atomic level, meaning that it can be lost or gained only in small discrete "packages." The mathematical expression for these small packages is the quantity $h \, v$, where h represents Planck's constant. Experimental evidence has established that Planck's Constant is equal to 6.626×10^{-34} joules second. Using this constant, physicists have determined that the energy (measured in joules) of an electromagnetic wave is described in the following equation:

$$E \text{ (energy)} = h \times c/\lambda$$

This leads to the idea that electromagnetic radiation is itself quantized. Einstein used this concept to show that electromagnetic radiation exhibited characteristics of particulate matter, a stream of particles called "photons," in addition to its well accepted wave properties. This has become known as the dual nature of light (Zumdhal, 1989). Furthering this research, Louis de Brolgie showed that the opposite was also true, namely, that matter can exhibit wave-like properties. From this, it was concluded that all matter has both particulate and wave properties, and that matter can be considered just another form of energy. Even large pieces of matter, which exhibit considerable particulate nature, can also be viewed as a dense concentration of waves. From this perspective our universe is a dense sea of electromagnetic waves with occasional concentrated areas which appear to us as matter.

In order to simplify things, scientists have classified electromagnetic waves according to their wavelength (or frequency) into what is commonly referred to as the electromagnetic spectrum. The portion of the electromagnetic spectrum typically displayed includes electromagnetic radiation of wavelengths ranging from 10^{11} mm to 10^{-11} mm (Lide, 1990) and includes the following divisions: radio waves, microwaves, infrared waves, visible light, ultraviolet waves, x-rays and gamma or cosmic rays. As the names of these divisions imply, knowledge of this segment of the electromagnetic spectrum has had tremendous impact on technological applications and has opened a whole new world of communication technology. For the most part, the other sections of the electromagnetic spectrum (i.e., below 10^{-11} mm and above 10^{11} mm) have been de-emphasized because no known technology has yet been discovered in these regions. However, as science and technology has progressed during the last century, the usable range of the electromagnetic spectrum has also expanded.

The Planck Length Defines the Shortest Wavelength in the Electromagnetic Spectrum

The complete range of the electromagnetic spectrum may be defined by two end points representing electromagnetic radiation of the shortest wavelength and electromagnetic radiation of the longest wavelength. According to quantum physics, the smallest possible dimension is described by what is known as the Planck Scale. At the Planck Scale all force fields and particles are believed to be completely united into a single unified field. From this fundamental level of nature's dynamics the entire physical universe is created. Beyond the level of the Planck Scale our classical notion of space and time becomes meaningless, and it is impossible to measure anything. In fact, at this level space and time do not exist, implying that our concept of measurement becomes meaningless. Based on experimental evidence, the physical dimensions of the Planck Scale are mathematically derived as follows: the Planck Length is equal to about 10^{-32} mm; the Planck Time is equal to about 10^{-43} seconds; and the energy at the Planck Scale is equivalent to about 10^{19} GeV/c^2. Therefore, electromagnetic radiation with a wavelength equal to the Planck length would represent the theoretical endpoint of the smallest wavelength for an electromagnetic wave. As we approach measurements in the order of the Planck length, the energies involved becomes so enormous that we would need advanced technology to harness the potential of these finer levels. For example, using the current technology of particle accelerators, it would require an accelerator with the circumference about the size of our galaxy to probe even to 10^{-28} mm.

The Size of the Universe Defines the Largest Wavelength in the Electromagnetic Spectrum

Intuitively, the largest possible wavelength for electromagnetic radiation would be confined to the size of the known universe. The usual equation used to calculate the diameter of our universe is derived from what is known as Hubble's law. This mathematical equation was formulated by the famous cosmologist, Edwin Hubble, who found that the velocity, v, of a galaxy moving away from an observer is proportional to how far away it is from us. This relationship is summarized in the following formula (Hubble, 1958), where d stands for distance, v for velocity and H for Hubble's Constant:

$$v = H \times d$$

This means that the farther away a galaxy is, the faster it is moving. It should be noted, however, that the actual numbers obtained from this equation are approximate because there is noticeable inaccuracy in the experimental measurements at this order of magnitude. By taking the reciprocal of Hubble's constant, it is possible to obtain the appropriate units from which we can determine the age of the universe. From such calculations, the universe is estimated to be about twenty billion years old. This number is consistent with other predictions obtained from the abundance of radioactive elements and the oldest globular clusters in the Milky Way galaxy.

Assuming that the edges of the universe expand at the speed of light, we can use the present age of the universe to calculate its diameter, which is equal to about 10^{29} mm. However, the exact size of the universe is not known, and this number may deviate by several orders of magnitude. Exact calculations are restricted due to inaccuracy in experimental measures and inherent limitations of current cosmological theories. For instance, some theories suggest that the value of Hubble's Constant may change according to the age of the universe, but this hypothesis is, at this stage, still speculation. For the purpose of the current discussion, it is adequate to use 10^{29} mm as the approximate diameter of the known universe and as an upper limit to the wavelength of electromagnetic radiation.

Conceptually there are several reasons why the radius of the known universe would define the largest possible measurement of length. First, according to Hubble's Law, the velocity of a galaxy moving away is proportional to its distance. Thus, the greater the distance, the faster things are moving until we reach a distance beyond which all objects are moving away faster than the speed of light. This distance, equal to the calculated diameter of our known universe, is significant because the only way to obtain information about distant galaxies is from the light traveling to us from these places. Far-off galaxies are only observable because the light and other electromagnetic waves emitted from the galaxy have been traveling through space for millions of years. What we observe today, therefore, is the condition of those galaxies millions of years ago. The farther away the galaxy is, the more time it will take for the light to reach us. However, if the galaxy is beyond the "edge" of our known universe it would be traveling faster than the speed of light and we simply would never receive any information of its existence.

Another conceptual notion is derived from Einstein's theory of relativity. According to Einstein, objects traveling at speeds approaching the velocity of light begin to change in length as well as their orientation in time. As a result, the measurement of length and time become almost meaningless for any object traveling at this speed. This implies that measurements at or beyond the diameter of the known universe are subject to these conditions because the "edge" of the known universe is traveling at the speed of light.

Locating the Constitution of the Universe in the Electromagnetic Spectrum

We have now provided a basic understanding of the formation of electromagnetic radiation and its organization into the well-known electromagnetic spectrum. In addition, we have defined the theoretical endpoints of the electromagnetic spectrum, between which resides the portion of this spectrum typically displayed. Based on this background information, we will now proceed to locate in the electromagnetic spectrum the structure of the Constitution of the Universe.

In the first level of unfoldment of the Constitution of the Universe, the state of unity, represented by "A-K," is elaborated into the eight syllables of first pada (refer to Figure 1). Correspondingly, it is possible to organize the full range of the electromagnetic spectrum into eight general categories. Table 4 displays the complete electromagnetic spectrum starting from the smallest wavelength (Planck length) and ending with the largest wavelength (diameter of the known universe). Between these two endpoints are listed all other wavelengths of the electromagnetic spectrum by consecutive increments in powers of ten. By listing the wavelengths of the electromagnetic spectrum in this fashion, we find there are about sixty-four numerical divisions (refer to Table 4). Within this comprehensive list of the electromagnetic spectrum, we propose that there are eight

Table 4. Summary of the Correspondence Between the Constitution of the Universe and the Electromagnetic Spectrum

Eight General Divisions of Electromagnetic Waves:

Unified Fields	Elementary Particles	Nuclear	Atomic & Molecular	Planetary	Stellar	Galactic	Universal

Sixty–Four Waves Spanning the Planck Length to the Diameter of the Universe:
(Wavelengths in mm)

10^{-32}	10^{-31}	10^{-30}	10^{-29}	10^{-28}	10^{-27}	10^{-26}	10^{-25}
10^{-24}	10^{-23}	10^{-22}	10^{-21}	10^{-20}	10^{-19}	10^{-18}	10^{-17}
10^{-16}	10^{-15}	10^{-14}	10^{-13}	10^{-12}	10^{-11}	10^{-10}	10^{-9}
10^{-8}	10^{-7}	10^{-6}	10^{-5}	10^{-4}	10^{-3}	10^{-2}	10^{-1}
10^{0}	10^{1}	10^{2}	10^{3}	10^{4}	10^{5}	10^{6}	10^{7}
10^{8}	10^{9}	10^{10}	10^{11}	10^{12}	10^{13}	10^{14}	10^{15}
10^{16}	10^{17}	10^{18}	10^{19}	10^{20}	10^{21}	10^{22}	10^{23}
10^{24}	10^{25}	10^{26}	10^{27}	10^{28}	10^{29}	10^{30}	10^{31}

Each "Wave" of the Electromagnetic Spectrum is Described in Three Ways:

- Energy, measured in joules   (Rishi Interpretation)
- Frequency, measured in hertz   (Devata Interpretation)
- Wavelength, measured in millimeters   (Chhandas Interpretation)

natural divisions matching the eight prakritis of the Constitution of the Universe as located in the Rig Veda. The grouping of the electromagnetic spectrum into eight general groups can be rationalized on the basis of wavelength, energy and frequency of the electromagnetic radiation that characterizes each category. These eight divisions are summarized in Table 4 and are briefly described as follows:

- The first division includes all electromagnetic radiation with wavelengths between the Planck length at 10^{-32} mm and 10^{-25} mm. Included in this range are electromagnetic waves that possess enormous energy and very high frequency. According to quantum physics, at 10^{-32} mm and 10^{-28} mm we have what are called super unification and grand unification, respectively. Super unification refers to a level of nature's dynamics where all the forces and elementary particles are completely unified. In grand unification all forces and particles (except gravity) are unified. Electromagnetic radiation and its properties still exist, just that they become increasingly indistinguishable from the other forces of nature. At the distance 10^{-32} mm, the wavelength of electromagnetic radiation approaches the distance of the Planck length, the smallest definable dimension beyond which measurement as we know it becomes meaningless.

- The second division includes all electromagnetic radiation with wavelengths between 10^{-24} mm and 10^{-17} mm. Included in this range are electromagnetic waves that possess very high energy levels and the size of the wavelengths are in the order of the elementary particles of physics. Although no known technology has been discovered for radiation in this portion of the spectrum, it is possible that future applications could make use of these high energy realms.

- The third division includes all electromagnetic radiation with wavelengths between 10^{-16} mm and 10^{-9} mm. The energy associated with the frequency of electromagnetic radiation that occurs in this range is in the order of nuclear energy, hence we have called this division of the electromagnetic spectrum the "nuclear" range. Included in this arena are gamma rays and cosmic radiation responsible for the high energy electromagnetic waves that enter the Earth's atmosphere from outer space. Gamma rays are also emitted by certain radioactive elements and are equivalent to several hundred thousand electron volts.

At the distance scale of about 10^{-15} mm, we have what is known as electroweak unification. This term is used in quantum physics to describe a partially unified field that includes the weak force and the electromagnetic field. At 10^{-15} mm these two fields are fundamentally indistinguishable from one another. In other words, the photon responsible for electromagnetism behaves the same as the massless $W^{\pm}$ and Z^0 bosons of the weak force. Only when the symmetry of the electroweak force is broken do the $W^{\pm}$ and Z^0 bosons become very massive and assume different properties.

- The fourth division includes all electromagnetic radiation with wavelengths between 10^{-8} mm and 10^{-1} mm. Within this range, the energy of electromagnetic radiation is in the order of the atomic and molecular levels. As a result, electromagnetic waves in this region are used typically in spectroscopy for compound identification and concentration of unknown substances. At 10^{-7} mm we have x-rays, at 10^{-5} mm ultraviolet radiation, between 4×10^{-4} mm and 7×10^{-4} mm visible light and at 10^{-1} mm infrared waves.

- The fifth division includes all electromagnetic radiation with wavelengths between 1 mm and 10^7 mm. Included in this range are microwaves at about 1 mm in wavelength, and FM and AM radio waves between about 10^2 mm and 10^7 mm in wavelength. Because these waves exhibit relatively low energy levels and the distance of the wavelengths are macroscopic in size, we have labeled this division of the electromagnetic spectrum as the "planetary" range.

- The sixth division includes all electromagnetic radiation with wavelengths between 10^8 mm and 10^{15} mm. Radio waves, long range power waves and AC circuits are located at the beginning of this portion of the spectrum . For electromagnetic radiation with wavelengths beyond 10^{12} mm there is as yet no known technological applications. Because of the large size of wavelengths for electromagnetic waves in this division, we have called this segment of the spectrum the "stellar range."

- The seventh and eighth divisions include all electromagnetic radiation with wavelengths between 10^{16} and 10^{23} mm and 10^{24} and 10^{32} mm, respectively. The size of these wavelengths is approaching galactic and universal dimensions, as suggested by the names we have assigned to them. As we approach 10^{32} mm (approximate diameter of the universe) the energy of electromagnetic radiation approaches zero. This is theoretically the endpoint for the largest possible electromagnetic wave. As technology advances, it is possible that electromagnetic radiation in these larger wavelengths will become increasingly significant.

We may now continue our connection with the Constitution of the Universe. From the eight basic syllables of the first pada of Rig Veda, the next stage in the expansion of the Constitution of the Universe is the interpretation of each of the eight syllables with respect to rishi (knower), devata (process of knowing) and chhandas (known). This produces a twenty-four-fold structure. Likewise, there are three basic aspects describing an electromagnetic wave that correspond to the interpretations of rishi, devata and chhandas. The first characteristic, the energy of an electromagnetic wave, matches the abstract and subtle nature of rishi. The second characteristic, the frequency of an electromagnetic wave, denotes the number of waves passing a particular point in a given time. This corresponds with the devata interpretation which embodies the qualities of active, moving and dynamic. The last characteristic, the wavelength of electromagnetic radiation, is the physical dimension of a wave and corresponds with the materialistic and concrete nature of chhandas. When viewed with respect to these three characteristics of an electromagnetic wave, the eight basic divisions of the entire electromagnetic spectrum expand to a total of twenty-four units, matching the twenty-four units of the Constitution of the Universe.

In the next level of elaboration in the Constitution of the Universe, the eight syllables in the first pada of Rig Veda expand into sixty-four units. We propose that this sixty-four-fold expansion matches the sixty-four numerical divisions of the entire electromagnetic spectrum (from the Planck length to the diameter of the universe). In Table 4 we have displayed these sixty-four divisions and have highlighted the middle one third, corresponding to the portion of the electromagnetic spectrum that is typically displayed and used in current technology. Although these sixty-four divisions are purely numerical and do not appear to correspond exactly to the various individual bands of known radiation, such as x-rays and radio waves, this numerical categorization in powers of 10 is often used (e.g., Lide, 1990). Perhaps there is some underlying reason or significance to the division of the electromagnetic spectrum into powers of 10 which is still to be discovered.

The final correspondence with the Constitution of the Universe stems from the sixty-four basic syllables of Rig Veda interpreted with respect to rishi, devata and chhandas, thus comprising a total of 192 units. Likewise the sixty-four numerical divisions of the electromagnetic spectrum may be viewed with respect to the energy, frequency and wavelength of electromagnetic radiation (thus a total of 192 units).

Speculations on a "Hubble Scale"

In conclusion, we would like to reflect, for a moment, on the conceptual implications of the correspondences presented in this section. During our discussion of the theoretical limits or endpoints for the electromagnetic spectrum, we described the Planck Scale and the diameter of the known universe. As we stated earlier, the Planck Scale is a level that has been established by quantum physicists as the smallest possible dimension, beyond which measurement is vague or meaningless. At present however, scientists have not established a scale that would define the largest possible dimension (i.e., the "reverse" or "opposite" of the Planck Scale). Intuitively we are led to believe that such a scale should exist. The obvious dimensions for such a scale would be the size of the known universe based on Hubble's Law. Using this information, we can define a "Hubble Scale" of the largest possible dimensions, beyond which measurement would be meaningless.

The length of this postulated "Hubble Scale" would be the diameter of the known universe which is about 10^{29} mm. According to Hubble's Law, the velocity of a galaxy moving away from us is proportional to its distance from us. As described earlier, this leads us to conclude that there exists a distance where objects must (appear to) be traveling away from us at the speed of light, and that this distance is the size of the known universe. In other words, the "edge" of our universe is traveling at the speed of light. Beyond this "edge" nothing would appear to exist because for any entity traveling faster than the speed of light, information about its existence would never reach us. Furthermore, based on Einstein's theory of relativity, if the "edge" of the universe is traveling at the speed of light, measurements approaching this distance will be subject to increasing time and length distortions. These reasons imply that the diameter of our universe may be a "Hubble length" that defines the largest measurement possible.

Another concept lending support to the idea of a "Hubble Scale" follows. One reason why measurement becomes ill-defined at the Planck Scale is because the intrinsic mass energies of quantum fluctuations become so large that gravity participates in a major way in quantum dynamics. Perhaps a similar process is also working at larger scales. At distance scales approaching the diameter of the universe, the effect of all the mass and gravity of the universe could conceivably exhibit an increasingly significant influence on experimental measurement to the degree that they become meaningless.

The idea of two universal scales (one for the smallest possible dimension and the other for the largest possible dimension) is supported by the wisdom of the Ancient world. For example, in ancient Vedic India the enlightened sages believed that "As is the microcosm, so is the macrocosm. As is the atom, so is the universe" and "Smaller than the smallest is the same as bigger than the biggest." Throughout the ancient world there is also the famous quote "As above, so below."

Conceptually what the idea of the "Hubble Scale" implies is that, wherever we are in the universe, we can imagine a sphere around us having the distance of the "Hubble length," forming a kind of "bubble" that defines our universe. Regardless of our location, we would always be at the center of "our" universe. Interestingly, this same viewpoint is accepted by modern cosmologists who believe that, regardless of our location in the universe, we will always observe the same phenomena; that is, in any direction we look all galaxies appear to be moving away. For example, if we could observe a space traveler moving away from us, we would see him moving out towards the far reaches of "our" universe. However, his viewpoint would be quite the opposite, he would see us moving towards the far reaches of "his" universe. Eventually, the space traveler would be moving away at the speed of light and would thus disappear from our view. In one sense, we could think of the cosmos as an ever-expanding universe. Essentially this means that no two people share exactly the same universe. Everyone is, in effect, their own universe. We leave further speculation on such ideas up to the reader's imagination.

Section III

Electronic Information

Computers will do exactly what we tell them to do. They are programmable electronic devices that can store, retrieve and process information. The value of electronic computers is their ability to represent and transform information in order to produce useful outputs from given inputs. In current computer technology, all information and data processing is based on the binary nature of electromagnetic interactions that provide an incredibly simple means of computing universally applicable data. In this section we will summarize the basic components of electronic computers and demonstrate that their design reflects the numerical structure inherent within the Constitution of the Universe.

Development and Fundamental Concepts of Computers

Historically, the first computing devices were the mechanical calculators manufactured during the 19th century. Through a complex array of gears and wheels these first calculators had the ability to perform basic arithmetic operations. By 1940 the first computer was built using what were called "electromechanical relay switches" which controlled the flow of electric current. A relay switch acted as the central computational device processing information according to changes in the flow of electric current. Following this invention, the relay switches were replaced by vacuum tubes that performed the same function as relay switches, but were faster and less expensive. Computers built on vacuum tubes, however, were large in size, weighed about thirty tons and required excessive power as well as constant upkeep. During the 1950s computer technology radically changed with the invention of a physical device called the transistor which replaced vacuum tubes. Since that time the tremendous growth in computer technology has been based on the speed and small size of transistors used in, what are known as, electronic circuit chips. As a result, modern-day computers have enormous computational speed, increased memory storage and are affordable. Today we are still in the midst of this revolution of computer-related applications and devices.

The transistor, one of the most important components of computer hardware, is able to regulate and utilize changes in the electromagnetic field. Generally, transistors are built using silicon or germanium which exhibit semiconducting properties. By adding small amounts of impurities such as boron or indium to a semiconducting material, free positive and negative charges are created. When voltage (electric potential) is applied to a computer circuit, it produces the flow of electric charge known as current. The tremendous value of the transistor lies in its ability to act as a switch which is turned "on" and "off" in response to changes in voltage. If the voltage is high, then a large current can flow with no resistance allowing the transistor to be turned "on." Conversely, if the voltage is low, the flow of current is blocked representing the "off" state of a transistor. Using transistors in combination with other electronic components, such as capacitors and resistors, it is possible to build the complex hierarchy of computer hardware.

Electronic Information and Memory Units are Based on Binary Numbers

In an electronic computer the "on" and "off" states of a transistor provide the two fundamental ways of recording information. These two basic conditions are represented by binary numbers. High voltage ("on" state of a transistor) corresponds to "1" and low voltage ("off" state of a transistor) signifies "0." Using these binary numbers, all information and data processing can be expressed in a computer system. For instance, the control of large scale electronic activity, such as displaying letters on a computer monitor or the turning on of lights, is governed by binary digits. Computers are "happy" with binary numbers because it is easier to design reliable physical devices which have only two possibilities: 1 and 0, "on" and "off," or "yes" and "no." Computer operations and computations are substantially simplified. As an interesting note, the original decision to use the binary number system of "on" and "off" functioning in computers was influenced by the "on" and "off" firing of neurons in the human brain. The field of cybernetics that studies and compares electronic computers to the human nervous system has arisen out of this initial work . Recently this research has expanded into the area of neural networks, using the knowledge of how nerve cells operate in the construction of new types of computing systems.

Binary numbers are base 2 which means that everytime we get the number 2 we carry or shift a digit one place to the left. In other words, each place to the left increases by a factor of 2. For example, in base 2 the number "10" is read "one zero," not "ten". Even though operations are longer with binary numbers, it is much easier for a computer. In the language of computer science, a binary digit (0 or 1) is the fundamental unit of memory for storing information and is called a BIT (abbreviated from Binary digIT).

The value of the BIT (0 or 1) is determined by the presence or absence of a voltage drop in a transistor. An alternative way to represent the value of a BIT is by the orientation of a small magnetic ring. When the ring is magnetized in one direction it represents "0" and when magnetized in the opposite direction it will represent "1." Changing the direction of the magnetism can be achieved by simply applying electrical current to the wire connected to the magnetic ring.

Using only one BIT, we are able to record two different types of data (the first represented by "0" and the second by "1"). With a set of two BITS we will be able to distinguish between four different pieces of information (i.e., "00," "01," "10" and "11"). In order to obtain enough different combinations to account for all numbers and letters, computers typically use a sequence of eight BITS or binary digits. For a sequence of eight digits (each digit has two possible values, "0" or "1") the total number of unique sequences is equal to 2^8 or 256. This adequately covers all numbers (0 through 9), all letters (lower case and upper case) and various other symbols and punctuations. For example, using a sequence of eight BITS, a computer may represent the letter "A" by the code "10100001" and the number "5" by "01010101." Likewise, all other letters, numbers and symbols would be represented by a unique array of eight binary digits.

An array of eight BITS is called a BYTE (also known as one Alphanumeric Character). These eight-BIT groupings are considered to be the funda-mental units of memory for most electronic computer systems. Although computer memory can be expressed in BYTES, the symbol "K" is some-times used when memory size is large. One "K" of memory is equal to 2^{10} or 1024 BYTES. Thus if a computer has 256 K of main storage it can hold 262,144 BYTES or characters of data in memory.

Groups of BYTES define an even larger division in computer memory, called the "word." Words are composed of a sequence of BYTES ranging from 2 to 8 (i.e., 16 to 64 BITS). This means that the size of each word is defined in terms of multiples of eight BITS. In modern computers most systems have eight BYTES (64 BITS) in each word. Each word is treated as a unit or location of memory which can be used to store information for processing or retrieval. The size of a word determines how much information it can hold.

All letters and numbers that we enter into a computer are translated or compiled into a sequence of binary digits. This binary-coded information is often called the "machine language" of computers. Machine language is used directly by the computer to store, process and retrieve information. However, if we were to communicate with a computer in machine language, the process would be very tedious and time-consuming. As a result computer scientists have developed what are known as "high-level" programming languages such as BASIC, PASCAL, FORTRAN and C that allow us to interact with computers in words and numbers which are familiar to us. When a computer receives instructions in a high-level programming language, the information is "compiled" or translated into machine language which can then be used by the computer to perform operations.

Electronic Information and the Constitution of the Universe

We have now arrived at a basic understanding of the essential components of electronic computers and foundational role of binary numbers in storing, processing and retrieving information. Using these principles, we may now investigate how the structure of the Constitution of the Universe can be located in the organization of computer information.

Our first comparison between electronic information and the Constitution of the Universe begins with the primal unit, represented in the Rig Veda as "A-K." According to the Vedic Literature, "A" represents complete knowledge and a state of all possibilities, which collapses to the point value of "K." The letter "K" represents a state of zero possibilities or emptiness. Likewise, all possible information and numbers in a computer are represented by the patterns of zeros and ones. The transformation of these patterns is the essence of all computations and computer operations. We propose that the letter "A" corresponds to the binary digit "1" representing the "on" state of a transistor or the presence of voltage. The fullness of the electromagnetic field is embodied in voltage matching the qualities associated with "A." We further propose that the letter "K" corresponds to the binary digit "0" representing the "off" state of a transistor or the absence of voltage.

The Constitution of the Universe subsequently unfolds the first syllable "A-K" into the eight syllables or eight prakritis of the first pada (refer to Figure 1). Similarly, in computer systems the fundamental unit of memory for storage and retrieval of information is a sequence of eight BITS (called a BYTE). The BYTE thus corresponds well to the first pada of eight syllables in Rig Veda.

Further elaboration of the Constitution of the Universe provides an interpretation of the eight syllables with respect to rishi (knower), devata (process of knowing) and chhandas (known). This produces a total of twenty-four syllables. Similarly, there are three basic areas of computer science corresponding to the nature of rishi, devata and chhandas. The first aspect, computer theory, consists of the mathematical and logical information involved in electronic technology and the formation of binary digits. This aspect of computer science matches with the abstract and refined qualities of rishi. The second aspect, computer software, is composed of the programs used to tell the computer what to do. As mentioned earlier, there are a variety of high-level programming languages to communicate with computers. Computers then translate the information we give them into machine language which is essentially just binary digits. These qualities of computer software match the dynamics and interactive nature of devata. The third aspect, computer hardware, includes the physical devices. The transistor which creates binary numbers by regulating the flow of electric current is one of the most important physical devices. This identically matches the physical and material nature of chhandas. Together these three foundational areas of computer science provide a full description of binary digits.

In the next level of elaboration of the Constitution of the Universe, the eight syllables in the first pada of Rig Veda expand into sixty-four units. Likewise, in the computer system BYTES are organized into larger units of memory storage called words. Words usually consist of two to eight BYTES or sixteen to sixty-four BITS (binary digits). The length of each word is thus a multiple of eight binary digits. Although the number of BITS in a word is not always sixty-four, we propose that, because many computers do use sixty-four BITS/word, this is a reasonable parallel with the Constitution of the Universe.

Finally within the Constitution of the Universe the sixty-four basic syllables of Rig Veda are interpreted with respect to rishi, devata and chhandas, thus comprising a total of 192 units. Similarly, all the binary digits which make up a word can be described through the three basic areas of computer science. The correspondences with the Constitution of the Universe presented in this section are summarized in Table 5.

As an ending note, it is worth mentioning two other independent authors who have found universal principles operating in the binary numbers. Dr. Schonberger published evidence in 1973 on the relationship of the binary number system to the genetic code of DNA and Jose Arguelles located in the Tzolkin or ancient Mayan calendar system a sequence of binary numbers (in 1984). Both the DNA and the Mayan calendar system are described in detail elsewhere in this book.

Table 5. **Summary of the Correspondence Between the Constitution of the Universe and Electronic Information**

---

1. BIT (Binary DigIT) – "A–K"

- A binary number (base two) has two possible values : "1" or "0"
- In computer memory, "0" represents "off" and "1" represents "on"
- Each binary digit is called a BIT and represents the smallest unit of data

---

2. BYTE (Alphanumeric Character) – Eight Prakritis

- Sequence of eight BITS, the fundamental unit of memory on which a computer operates

---

3. Word – 64 Syllables in Rig Veda

- A word is composed of a sequence of BYTES, usually ranging from 2 to 8 (16 to 64 BITS).

---

4. Three Foundational Areas of Computer Science

- Computer Theory   (Rishi Interpretation)
- Computer Software   (Devata Interpretation)
- Computer Hardware   (Chhandas Interpretation)

CHAPTER 3:

CHEMISTRY AND THE CONSTITUTION OF THE UNIVERSE

SECTION I

THE PERIODIC TABLE OF ELEMENTS

In this section, we will discuss the current understanding of atomic structure that is the foundation for modern chemistry. Based on this information, chemists have been able to develop what is known as the periodic table of elements, which has greatly aided in the understanding of the structure and interconversions of matter. Further, we will demonstrate that hidden within the organizational structure of the atom is the numerical pattern of the Constitution of the Universe.

Historical Perspective

The language of chemistry, or the study of matter, begins with the element. An element is defined as a fundamental substance which cannot be broken down into simpler substances by chemical or physical means. Several common examples of elements include: oxygen, nitrogen, sodium, zinc and iron. "Pure" elements in nature, however, are uncommon. They often combine with one another to form compounds (e.g., water which consists of oxygen and hydrogen) or mixtures (e.g., air is a mixture of oxygen, nitrogen and other gases).

Our current list of known elements developed over a period of several centuries. Prior to the turn of the nineteenth century, relatively few elements were known to exist and they were recorded simply as a list of substances, lacking any logical order. However, during the nineteenth century there was a proliferation of newly discovered elements. By 1830 there were fifty-five recognized elements, and scientists began to search for an orderly arrangement to classify them (Asimov, 1965). One of the ways employed to organize the elements was the order of increasing atomic weights. The atomic weight of an element is the average mass of its atoms. Among the first to use this approach was the English chemist, John Newlands (1837 to 1898). After arranging the elements known at that time according to their atomic masses, Newlands noted that particular chemical characteristics were found to repeat every eighth element (similar to the octaves of music).

Newland's ideas, however, were not generally accepted until they were elaborated by the famous Russian chemist, Dmitri Mendeleev (1834 to 1907). As with Newlands, Mendeleev noticed that some elements exhibited remarkable similarity in their chemical behavior. On this basis, he constructed one of the first "period tables of elements" in such a way that elements in the same vertical column of the table possessed similar chemical properties. Upon completion of the table, Mendeleev found that there were a number of gaps in the sequence of the elements. However, instead of viewing these gaps as imperfections, he proposed that they were elements yet undiscovered. Shortly afterward, these predictions were dramatically confirmed to amazing closeness. As a result, Mendeleev's table became universally accepted as an essential guide and a source of tremendous information for chemists. Currently, the number of elements in the periodic table is 110. Ninety of these elements are found naturally occurring in our environment. The remaining twenty have been synthesized by scientists in the laboratory, but are not known otherwise to exist on Earth.

Atoms are Composed of Electrons, Protons and Neutrons

Structurally, all elements are composed of tiny particles called atoms, which are fundamental units of matter consisting of a dense, positively charged nucleus surrounded by a cloud of negatively charged electrons. An element consists of a conglomeration of identical and unique atoms that are distinct from the atoms found in all other elements. Each atom consists of three basic components: protons, neutrons and electrons. The nucleus or center of the atom contains protons, which have a positive charge, and neutrons, which have no charge yet possess the same mass as protons. Compared with the overall size of the atom, the nucleus is small and has an extremely high density. Revolving around the nucleus are the electrons which carry a fixed negative charge, equal in magnitude to the positive charge of a proton, and which have a mass about 2,000 times less than either the proton or the neutron.

The total number of protons in the nucleus of an atom is called the atomic number and this is used to identify an element. In the periodic table used by chemists today, the elements are organized according to their atomic numbers in ascending order. Although this organization is different from the original periodic table of Mendeleev, which was based on atomic weights, the differences between the actual sequence of elements in the two tables are slight.

Typically, elements are electrically neutral, thus, the number of positively charged protons is equal to the number of negatively charged electrons. However, under various circumstances an element can lose or gain electrons, but in either case the name of the element does not change. A gain of electrons will cause the element to have a net negative charge and is called an "anion." A loss of electrons causes the element to have a net positive charge and is called a "cation." The number of neutrons may also vary in an atom, but, because neutrons have no charge, the net charge of the atom will be unaffected by the gain or loss of neutrons. An atom containing an unequal number of protons and neutrons is called an isotope. Often isotopes are radioactive and have been used for many technological purposes. Regardless of the number of neutrons its nucleus contains, any one atom still exhibits the same chemical nature. If, however, the number of protons changes, then the name of the element changes. The number of protons in the nucleus specifies the distinct type of atom and hence identifies an element.

A Conceptual Understanding of Atomic Structure

One question which may arise is that if all atoms contain electrons, protons and neutrons, then why do different atoms have different chemical characteristics? The key to answering this question lies in the number and arrangement of electrons in the atom because electrons occupy most of the atomic volume and are thus the components that interact with other atoms to form molecules and all other types of matter. In fact, most of the chemistry of an atom, chemical bonding and chemical kinetics, mostly arise from the number and arrangement of its electrons. Further, the periodicity and many of the trends of the periodic table become evident from understanding the distribution and arrangement of electrons in an atom.

In order to understand the distribution of electrons, we first need to comprehend some basic principles of atomic structure. One of the first advances in this field came from the German physicist, Max Planck, who found that at the atomic level energy is "quantized," implying that energy can be gained or lost only in small discrete "packages." This also means that the energy of an electron is quantized and that in an atom electrons can move around the nucleus only in well defined energy-levels. This can be experimentally verified by studying the light emission spectrum of an element. When electrons are excited in an atom they contain excess energy, which they release by emitting various wavelengths of light. These discrete wavelengths produce what is known as an "emission spectrum" containing only a few lines of colored light. This indicates that the energy of the electron in an atom is quantized. If the energy of an electron were not quantized, the emission spectrum would look the same as white light – a continuous spectrum that contains all wavelengths and not just a few discrete ones.

The next key link in comprehending atomic structure came from Albert Einstein who concluded that electromagnetic radiation possesses a dual nature, that is, it has both a wave-like and a particulate nature. Extending these findings, Louis de Broglie demonstrated that the reverse was also true – that matter, in addition to being particulate, also exhibits wave properties. In other words, an electron is not just a small particle moving in space, but it can also be thought of as exhibiting the behavior of a standing or stationary wave which is spread out over a large space. Based on these ideas, Werner Heisenberg, Louis de Broglie and Erwin Schrodinger developed what is known as the wave mechanical model of the atom. The great success of this model has been its ability to explain the observed trends and periodicity of the elements in the periodic table. In this model, electrons are described by a wave function which is gener- ally called an "orbital." Viewing an electron as an orbital or wave function is consistent with a certain law in physics, the "Heisenberg uncertainty principle," which states that it is impossible to know simultaneously both the location and the momentum of an electron. However, based on the square of the wave function, we can construct a three-dimensional map of the probability of locating or finding an electron at a particular point in time and in space. These maps are used to define the orbital shape in which the electron is most likely to be found 90% of the time.

Using these basic ideas, we can proceed to describe the configuration or arrangement of electrons in an atom. There are three important terms used to describe the atomic distribution of electrons: quantum level, subshell and orbital. A quantum level is a well-defined energy-level, in which a number of subshells reside. The higher the quantum level, the greater the number of subshells. A subshell is a set of orbitals having a particular shape. Each orbital can contain two electrons.

Electrons Have a Defined Arrangement and Configuration

Figure 4 provides a conceptual diagram of how electrons are distributed in an atom. To begin, we will describe the various types of subshells. The first one, called the "s-subshell," has the shape of a sphere. This means that any electrons located in this subshell will, by statistical probability, reside most of the time somewhere within a spherical boundary. The shape of the sphere depicts the actual form of the three-dimensional wave of the electron for this subshell. Each "s-subshell" has one orbital containing a total of two electrons.

The second subshell, called the "p-subshell," has the shape of a long balloon squeezed together in the middle. As before, this shape represents the statistical probability where the electrons in this subshell will be located most of the time. This subshell has three orbitals (x, y and z directions in space) which can contain a total of six electrons (two electrons per orbital).

The third and higher subshells contain increasingly more complex shapes which are too technical to be discussed in this book. For the present purpose, it will be enough simply to mention these higher subshells and the total number of orbitals and electrons each is capable of containing. The third subshell, called the "d-subshell," has five orbitals which can therefore contain a total of ten electrons. The fourth subshell, called the "f-subshell," has seven orbitals containing a total of fourteen electrons.

Figure 4. The Arrangement and Distribution of Electrons in an Atom

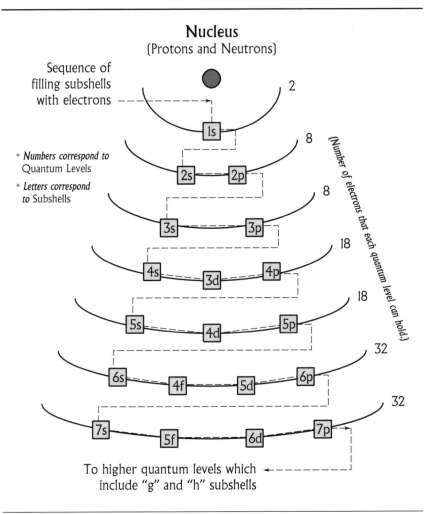

To higher quantum levels which
include "g" and "h" subshells

Maximum Number of Electrons per Subshell

- "s" subshell = 2   • "f" subshell = 14
- "p" subshell = 6   • "g" subshell = 18
- "d" subshell = 10   • "h" subshell = 22

Total Number of Electrons in "s" and "p" = **8**
Total Number of Electrons in "d", "f", "g" and "h" = **64**

The fifth subshell, called the "g-subshell," has nine orbitals containing a total of eighteen electrons. The last subshell to be considered is called the "h-subshell" and has eleven orbitals or a total of twenty-two electrons.

The order in which subshells are arranged in the quantum levels is summarized in Figure 4. As the quantum levels increase, the number of subshells located in each level also increases according to the order of subshells (i.e., "s," "p," "d," etc.). For the first quantum level there is one subshell, 1s-subshell. For the second quantum level there are two subshells: 2s-subshell and 2p-subshell. Likewise for higher quantum levels more subshells are added one by one. Thus, for each new quantum level we start by adding a new s-subshell, then a new p-subshell and so forth. However, in order to avoid ambiguous naming, the subshells for each quantum level are denoted by the number of that quantum level. (Note: in the higher quantum levels shown in Figure 4 certain subshells are filled prior to filling all the subshells from the previous quantum level. For example, in the fourth quantum level the order in which the electrons fill up the subshells is as follows: 4s → 3d → 4p [4s is filled prior to 3d because the 4s orbitals have a lower energy than the 3d orbitals]. In higher quantum levels the sequence in which the subshells are filled with electrons becomes increasingly more complex.)

Electrons fill the quantum levels and subshells starting from the first quantum level and proceed to higher levels according to the number of electrons in the atom. The sequence of filling up the levels is given in Figure 4 by the dashed line. Electrons are arranged in this sequence because the closer the electrons are to the nucleus of the atom, the greater the stability of the atom. It is important to remember that, regardless of the number of electrons in an atom, all the quantum levels and subshells are always present even if they are not occupied by electrons. If electrons are excited enough they temporarily "jump" into higher orbitals.

The Location of an Electron is Specified Using Quantum Numbers

To describe the location of any particular electron, we need to use four different quantum numbers. The first quantum number, called the "principle quantum number" (n), can have the value of 1, 2, 3, etc. Each of these numbers refers to the quantum level on which the electron resides.

The principle quantum number is also used to describe the energy and size of the orbital. The second quantum number, called the "azimuthal or orbital quantum number" (l), denotes the subshell the electron is occupying (e.g., s-subshell, p-subshell, etc.). The third quantum number, called the "magnetic quantum number," is the physical orientation of the subshell where the electron is located. In the s-subshell there is only one orientation because it is the shape of a sphere. For the p-subshell, however, there are three possible orientations (x, y and z axes of space). Each of the higher subshells have many different possible orientations which determine the total number of orbitals permitted for that subshell. The fourth quantum number is called the "electron spin quantum number" (M_s). A given electron can spin in one of two opposite directions, spin-up (+1/2) or spin-down (-1/2) (see Chapter 10 for detailed discussion of spin-types). Because each electron is electrically charged, when it spins it produces a magnetic field. Thus, the two types of spin will produce two oppositely directed magnetic fields.

The main significance of the quantum numbers is to locate accurately any electron. According to a postulate known as the "Pauli exclusion principle," in a particular atom no two electrons can have the same set of four quantum numbers. Since each orbital can hold two electrons, this means that the two electrons must have opposite spins.

The Periodic Table of Elements

At this stage, we are now able to use our understanding of the arrangement or configuration of electrons in an atom to grasp the organization of the periodic table. Before proceeding, there are two new terms which need to be defined: valence electrons and core electrons. Valence electrons are the electrons residing in the outermost quantum level of an atom. These electrons in the outermost level are involved in chemical interactions and chemical bonding. The quantum level in which the valence electrons reside is called the valence quantum level. All the other electrons occupying the lower quantum levels are referred to as the core electrons.

Figure 5 presents a copy of the standard periodic table of elements. Each vertical column of the table is called a group or family. The elements within a specific group all have similar chemical properties. Each horizontal row is called a period. The elements in each row are determined in

Figure 5. Locating the Duet Rule, Octet Rule and Expanded Octet in the Periodic Table of Elements

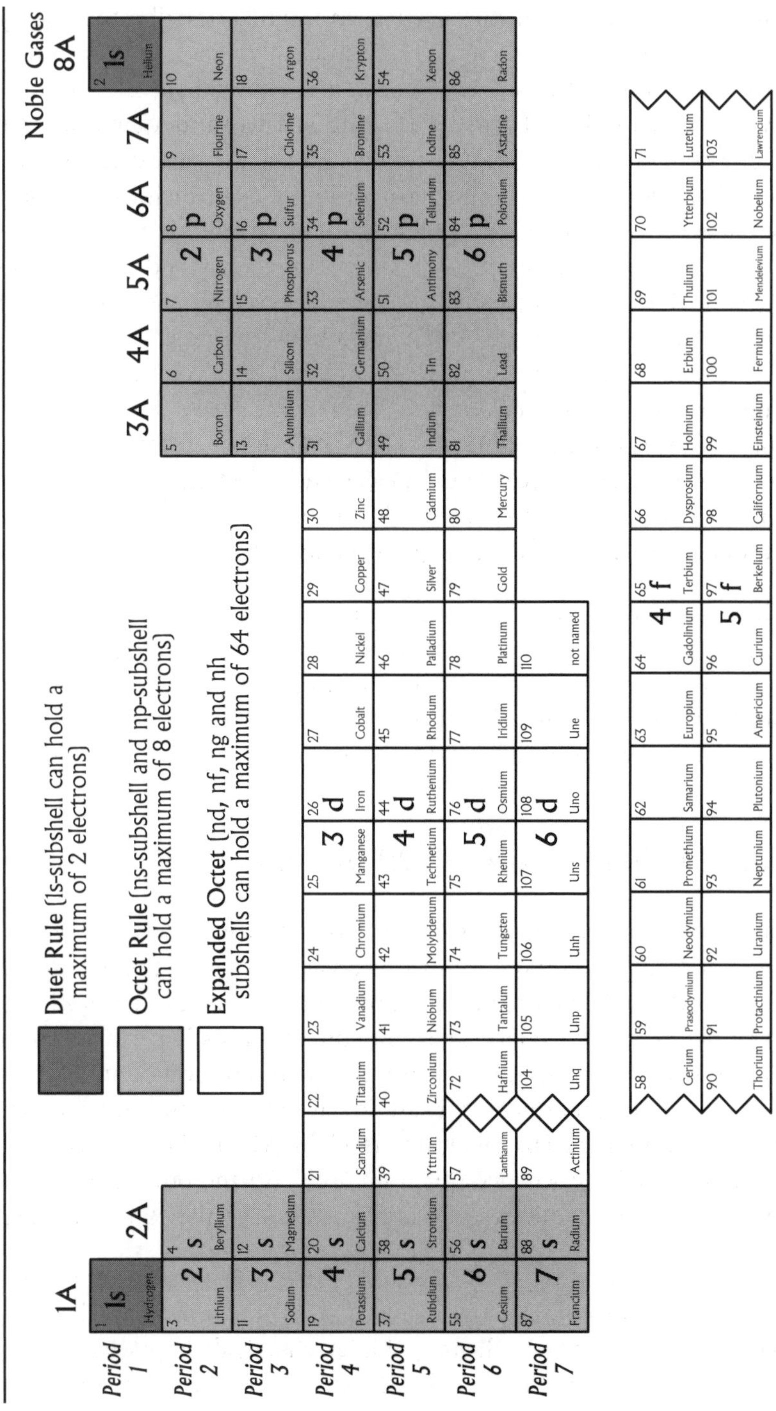

Duet Rule (1s-subshell can hold a maximum of 2 electrons)

Octet Rule (ns-subshell and np-subshell can hold a maximum of 8 electrons)

Expanded Octet (nd, nf, ng and nh subshells can hold a maximum of 64 electrons)

Period	1A	2A											3A	4A	5A	6A	7A	Noble Gases 8A
Period 1	1 1s Hydrogen																	2 1s Helium
Period 2	3 2 s Lithium	4 s Beryllium											5 Boron	6 Carbon	7 Nitrogen 2 p	8 Oxygen	9 Flourine	10 Neon
Period 3	11 3 s Sodium	12 s Magnesium											13 Aluminium	14 Silicon	15 Phosphorus 3 p	16 Sulfur	17 Chlorine	18 Argon
Period 4	19 4 s Potassium	20 s Calcium	21 Scandium	22 Titanium	23 Vanadium	24 Chromium 3 d	25 Manganese	26 Iron	27 Cobalt	28 Nickel	29 Copper	30 Zinc	31 Gallium	32 Germanium	33 Arsenic 4 p	34 Selenium	35 Bromine	36 Krypton
Period 5	37 5 s Rubidium	38 s Strontium	39 Yttrium	40 Zirconium	41 Niobium	42 Molybdenum 4 d	43 Technetium	44 Ruthenium	45 Rhodium	46 Palladium	47 Silver	48 Cadmium	49 Indium	50 Tin	51 Antimony 5 p	52 Tellurium	53 Iodine	54 Xenon
Period 6	55 6 s Cesium	56 s Barium	57 Lanthanum	72 Hafnium	73 Tantalum	74 Tungsten 5 d	75 Rhenium	76 Osmium	77 Iridium	78 Platinum	79 Gold	80 Mercury	81 Thallium	82 Lead	83 Bismuth 6 p	84 Polonium	85 Astatine	86 Radon
Period 7	87 7 s Francium	88 s Radium	89 Actinium	104 Unq	105 Unp	106 Unh 6 d	107 Uns	108 Uno	109 Une	110 not named								

58 Cerium	59 Praseodymium	60 Neodymium	61 Promethium	62 Samarium	63 Europium	64 Gadolinium	65 Terbium 4 f	66 Dysprosium	67 Holmium	68 Erbium	69 Thulium	70 Ytterbium	71 Lutetium
90 Thorium	91 Protactinium	92 Uranium	93 Neptunium	94 Plutonium	95 Americium	96 Curium	97 Berkelium 5 f	98 Californium	99 Einsteinium	100 Fermium	101 Mendelevium	102 Nobelium	103 Lawrencium

relation to the quantum levels. This means that period one will contain only those elements having their valence electrons existing in the first quantum level. There are only two such elements: hydrogen and helium. Period two contains only those elements with valence electrons existing in the second quantum level (i.e., lithium through neon). Likewise each period will contain only those elements with valence electrons existing in the corresponding quantum level.

As a quick reference or shorthand notation, chemists usually describe the location and number of electrons in an atom using what is called the electron configuration. For example, the electron configuration for first element, hydrogen, is "$1s^1$." This means that hydrogen has one electron (indicated by the superscript "1") located in the s-subshell of the first quantum level. The electron configuration of helium is "$1s^2$." This means that helium has two electrons located in the s-subshell of the first quantum level. Another more complex example is the electron configuration for sodium which is as follows: $1s^2\ 2s^2\ 2p^6\ 3s^1$. Likewise for any element it is possible to write an electron configuration.

The chemical reactivity or stability of an element is determined by how many valence electrons there are in the outermost quantum level. As a general rule, the most stable and unreactive elements have all the orbitals in their valence quantum level completely filled. On the periodic table these elements are called the noble gases and are located in column 8A of Figure 5. These include helium, neon, argon, krypton, xenon and radon. All other elements tend to "want" to gain or to lose electrons in order to achieve the electron configuration of a noble gas. This is especially true of elements in columns 1A and 7A (refer to Figure 5). Both these groups of elements are especially reactive and unstable because they need to gain or lose only one electron in order to achieve the electron configuration of a noble gas.

The "Duet Rule" and "Octet Rule"

From experimental evidence, chemists have demonstrated that the most important factor in the formation of compounds is for the atoms to achieve the configuration of a noble gas or a valence quantum level in which all the orbitals are completely filled with electrons. Several rules of chemistry have emerged based on these principles. The first is called the "duet rule" which applies only to the first quantum level. This rules states that all elements with valence electrons in the first quantum level have a tendency to have an electron configuration of helium, the first noble gas. In order to achieve the electron configuration of helium, the element needs to have a total of two (hence the term "duet" rule) electrons in the first quantum level which would fill the 1s-subshell. This rule only applies to the element hydrogen. Adding one additional electron to the hydrogen atom creates the electron configuration of helium.

For all elements beyond helium, we typically use what is called the "octet rule," originally conceived by G.N. Lewis in 1902. This rule is based on the tendency of elements to fill all orbitals completely in s-subshell and p-subshell in the valence quantum level. The total number of electrons that can fill an s-subshell and a p-subshell is equal to eight (hence the term "octet rule"). Thus the stability of an atom can be defined as a complete set of eight electrons in the outermost shell. When elements form compounds by covalent chemical bonding, they "share" electrons with each other in such a way that all the elements in the compound achieve the electron configuration of a noble gas. As a result, each element will react only with certain other elements in the formation of new compounds. For example, all elements in column 7A need one electron to achieve the electron configuration of a noble gas, and all elements in column 1A lack one electron. As a result, elements of these two groups readily interact with each other to form compounds. A classic example is the combination of sodium (element 11) and chlorine (element 17) to form sodium chloride, commonly known as table salt. Figure 5 summarizes elements that are generally affected by the duet and octet rules.

For the larger elements, it is often necessary to extend the octet rule to include the subshells higher than the s-subshell and the p-subshell. The total number of electrons which can fill the d-subshell, f-subshell, g-subshell and h-subshell is equal to sixty-four. (The numbers of orbitals and electrons which can fill each individual subshell were tabulated earlier.) For the larger elements, these higher subshells begin to be filled, affecting their chemical properties and behaviors. It needs to be noted, however, that no elements have yet been discovered with electrons in the g-subshell and h-subshell. Nevertheless, these subshells do exist and may be occupied temporarily by an electron if it is excited. At this stage, we can only speculate that more elements will be discovered and that the significance of a fully occupied h-subshell could be of even greater importance. It is interesting to note that some chemists believe that "superheavy" elements may exist with reasonable stability compared to the elements found in nature and those synthesized in the laboratory.

Atomic Structure Matches the Constitution of the Universe

So far we have discussed some of the basic principles of atomic structure and the electron configuration of atoms, providing the foundation for understanding the periodic table and chemical interactions. Using these principles, we will now proceed to locate in the arrangement and distribution of electrons in an atom the numerical structure of the Constitution of the Universe as located in Rig Veda. Our first correspondence relates to the first unit of the Constitution of the Universe, "A-K." According to Rig Veda, "A" represents the state of all possibilities, which collapses to the point value of "K." Likewise, the first element of the periodic table, hydrogen with its one electron, arises from unmanifest energy fields which contain the potential for all possible energy levels and subshells. Contained in the hydrogen atom are all possible quantum levels and subshells which become filled as new elements are created. Interestingly, hydrogen is the most abundant element in the universe. It makes up ninety percent of all atoms and three quarters of the mass of the universe (Lide, 1990). Further, all heavier elements are traditionally believed to have originated from hydrogen and helium via nuclear reactions in stars. Hydrogen is often used in suns (stars) as nuclear fuel: the "burning" of hydrogen generates heat and light, from which all biological life is nourished.

The next level in the Constitution of the Universe is the unfoldment of the first syllable, "A-K" into the eight syllables or prakritis of the first pada (refer to Figure 1). Likewise, in the periodic table chemists have developed what is known as the "octet rule," based on the understanding that the total number of electrons filling the first two subshells (s-subshell and p-subshell) is exactly equal to eight. As described earlier, the most important requirement for the formation of stable compounds is that the atoms achieve the configuration of a noble gas. Thus, elements will have a tendency to have all orbitals of the s-subshell and p-subshell completely filled in their valence quantum level.

From these eight syllables of the first pada of Rig Veda, the next level of expansion in the Constitution of the Universe is the interpretation of each of these eight syllables with respect to rishi (knower), devata (process of knowing) and chhandas (known). This produces a total of twenty-four units. Similarly, according to the 'Pauli exclusion principle," the location of any electron in an atom can be uniquely described using four quantum numbers. Three of these correspond to rishi, devata and chhandas. The rishi aspect embodies the qualities of abstract intelligence and energy. We propose that this corresponds to the azimuthal or orbital quantum number (l) which describes the shape and energy level of where the electron is located. The second interpretation in the Constitution of the Universe template is devata which embodies the qualities of active, moving and dynamic. This corresponds to the electron spin quantum number (M_s). As described earlier, each electron can spin or move in one of two opposite directions: spin-up (+1/2) or spin-down (-1/2). Since a spinning charge produces a magnetic field, the two types of electron spin create two opposing magnetic fields. The third aspect or flavor, chhandas, is equated with the material or physical properties of an entity. This corresponds to the magnetic quantum number (M_l) which relates to the physical orientation of the subshells where the electrons reside. Viewed with respect to these three quantum numbers, the group of eight electrons needed to fill completely an s-subshell and a p-subshell, make up a total of twenty-four units corresponding to the twenty-four units of the Constitution of the Universe. The fourth quantum number is the principle quantum number (n), which describes the size and energy of the electron orbital. We propose that this quantum number matches what is known in Rig Veda as the samhita value or the summation of rishi, devata and chhandas. The principle quantum number refers to a quantum level. The quantum level defines the basic division in the atomic structure. Within each quantum level we find the other three quantum numbers functioning to define the precise location of each electron.

For the next stage of elaboration in the Constitution of the Universe, the eight syllables in the first pada of Rig Veda expand into sixty-four units. We propose that this corresponds to what is known as the "expanded octet" in the periodic table. For the larger elements, the octet rule is often insufficient and higher subshells need to be taken into consideration as the number of the quantum levels increases. It turns out that the total number of electrons needed to fill completely the d-subshell, f-subshell, g-subshell and h-subshell is equal to exactly sixty-four which corresponds precisely with the sixty-four units of the expanded phase of the Constitution of the Universe.

The final correspondence with the Constitution of the Universe stems from the sixty-four basic syllables of Rig Veda interpreted with respect to rishi, devata and chhandas, thus comprising a total of 192 units. Likewise, we can view each of the sixty-four electrons needed to fill the higher subshells in the atomic structure of an atom with respect to the quantum numbers which define the exact location of each electron (thus a total of 192 units).

A Recapitulation

A summary of the correlations between the atomic structure of elements in the periodic table and the Constitution of the Universe is presented in Table 6. Another way to view these correspondences is to compare the sequential unfoldment of the Constitution of the Universe and the orderly hierarchy of subshells in the quantum levels. The numerical structure of the Constitution of the Universe unfolds in a particular sequence beginning with the first syllable, "A-K." From "A-K" the eight prakritis are created which in turn elaborate into sixty-four syllables within the first sukta. A very close structure is also present in the electron configuration of atoms. The electrons are added to orbitals in a sequence beginning with the first quantum level and proceeding upwards. At each new quantum level there are additional subshells added. For the first quantum level there is only the s-subshell containing only 1 orbital, or a maximum of two electrons. Once the first quantum level is filled, then electrons begin to occupy the second quantum level containing both an s-subshell and a p-subshell. The total number of electrons needed to fill both these subshells is exactly eight (matching the eight prakritis). In a similar fashion, at each new quantum level additional subshells are added. For quantum levels containing the d-subshell through the h-subshell, the total number of electrons needed to occupy all these orbitals is exactly sixty-four (corresponding to the sixty-four syllables within the first sukta of Rig Veda).

In summary, we have provided convincing evidence that the distribution of electrons in the atomic structure of elements perfectly reflects the numerical structure of unfoldment of the Constitution of the Universe. Further, it is interesting that chemists chose the terminology "octet rule" and "expanded octet" to describe the periodic table. Both these terms precisely correspond to the key numbers eight and sixty-four in the Constitution of the Universe. Understanding the electron configuration of the elements has allowed chemists to comprehend the trends in the periodic table and the mechanisms of chemical interactions.

Table 6. Summary of the Correspondence Between the Constitution of the Universe and the Periodic Table of Elements

---

Octet Rule:
(total number electrons in "s" and "p" subshells = 8; n refers to the quantum level or principle quantum number)

ns^1	ns^2	np^1	np^2	np^3	np^4	np^5	np^6

Expanded Octet:
(total number of electrons in "d," "f," "g" and "h" subshells = 64; n refers to the quantum level or principle quantum number)

nd^1	nd^2	nd^3	nd^4	nd^5	nd^6	nd^7	nd^8
nd^9	nd^{10}	nf^1	nf^2	nf^3	nf^4	nf^5	nf^6
nf^7	nf^8	nf^9	nf^{10}	nf^{11}	nf^{12}	nf^{13}	nf^{14}
ng^1	ng^2	ng^3	ng^4	ng^5	ng^6	ng^7	ng^8
ng^9	ng^{10}	ng^{11}	ng^{12}	ng^{13}	ng^{14}	ng^{15}	ng^{16}
ng^{17}	ng^{18}	nh^1	nh^2	nh^3	nh^4	nh^5	nh^6
nh^7	nh^8	nh^9	nh^{10}	nh^{11}	nh^{12}	nh^{13}	nh^{14}
nh^{15}	nh^{16}	nh^{17}	nh^{18}	nh^{19}	nh^{20}	nh^{21}	nh^{22}

Location of Each Electron is Uniquely Described by Four Quantum Numbers:

- Principle Quantum Number – n *(Samhita)*
- Azimuthal or Orbital Quantum Number – l *(Rishi Interpretation)*
- Electron Spin Quantum Number – M_s *(Devata Interpretation)*
- Magnetic Quantum Number – M_l *(Chhandas Interpretation)*

THE PRIMITIVE UNIT CELL AND CRYSTAL LATTICE

Throughout the ages, the intrinsic beauty of crystals has been admired by humankind. Only recently, however, has the study of the structure and properties of crystals emerged into the science we know as crystallography. As a multidisciplinary science, it has assembled knowledge from fields such as chemistry, physics, geology, thermodynamics and optics. When we think of crystals, the first images that come to mind are probably gems, diamonds or semi-precious stones. However, virtually all solids are crystals; in fact, the entire solid crust of the Earth is crystalline. In this section we will discuss the fundamentals of crystallography and provide convincing evidence that the basic structure of crystals and crystal growth reflects the numerical structure of the Constitution of the Universe.

Crystals Exist Primarily in the Solid State of Matter

A crystal (from the Greek word "krystallos," meaning "clear ice") is commonly defined as a "solid composed of atoms arranged in an orderly repetitive array" (Wood, 1964). Atoms, as described in the previous section, are fundamental units of matter consisting of a dense, positively charged nucleus surrounded by a constellation of electrons. What determines whether atoms are arranged in an orderly array, such as in crystals, is the physical state of the matter in which the atoms reside (Azaroff, 1960). There are three distinct physical states: gas, liquid and solid. In the gaseous phase atoms move independently of one another and as a result there is disorder and entropy in the arrangement of atoms. Such properties make it impossible for the formation of crystals in the gaseous phase of a particular substance.

In the liquid phase, atoms are closely packed (compared with the gaseous phase) restricting the motion of individual atoms. For example, at ordinary atmospheric pressure and temperature a given number of atoms in a gaseous phase occupies approximately 1,000 times the volume of an equal number of atoms in the liquid or solid state. Although atoms are so much closer together in the liquid state than in the gaseous, there is still some disorder and entropy in the arrangement of the atoms. Atoms are able to move past one another very easily, which is the reason why liquid can "flow." There are, however, attractive forces between the atoms in the liquid state which are essentially non-existent in the gaseous phase. For example, when you spill liquid mercury, tiny drops or beads scatter everywhere. The drops are spherical because of the internal attractive forces between the mercury atoms. Until recently only matter in the solid state was considered capable of forming crystals. However, evidence now indicates that under certain conditions "liquid crystals" are possible which have properties of being a liquid yet simultaneously also crystalline (i.e., an orderly arrangement of atoms). Such "liquid crystals" have already become useful in applied technology such as in the building of watches.

The last of the three states of matter, the solid state, results when a liquid is cooled enough so that it freezes. Generally when a liquid freezes, it becomes more viscous and there is a slight compression (with a few exceptions such as water). Atoms in a solid state are close enough so that they are unable to move past one another as they can in the liquid state. These atoms have a fixed and average position providing the foundation for an orderly arrangement. This orderly atomic arrangement defines a crystal, in which a specific pattern is repeated again and again. It is analogous to a hotel with floor upon floor of equivalent rooms and each room is furnished with the same items. Although virtually all solids are classified as crystals, there are a few exceptions which may result from the process by which crystals are formed. As will be discussed later, one of the most common methods of crystal formation is the cooling of a liquid to a solid. If the cooling process is relatively slow, then the atoms will assume an orderly arrangement representative of a crystal. However, if the cooling process is abrupt, then an amorphous solid or glass may result that does not have the characteristics of a crystal (e.g., volcanic glass).

Interestingly, although the solid state of matter exhibits the greatest degree of order in the arrangement of atoms, it also exhibits the greatest diversity. Each solid has its own preferred orderly arrangement of atoms giving rise to myriad elements, rocks and minerals. The differences between liquids, however, are less pronounced, but there is still some diversity (e.g., the inability of oil to mix with water). Gases, on the other hand, are very similar and there is not so much diversity. For example, they all readily mix and dissolve in one another (Holden and Morrison, 1982).

The Primitive Unit Cell is the
Building Block of Crystal Structures

Much of our current information on the structure and internal arrangement of atoms in crystals has been made possible because of x-ray diffraction studies. After the discovery of x-rays in 1895 by Wilhelm Roentgen,
it was found that a crystal will scatter a beam of x-rays in all directions and
into a large number of individual rays. The pattern of the scattered rays
provides knowledge of the arrangement of atoms in the crystal and the
distances between neighboring atoms. Various formulae such as Braggs
equation (discovered by Sir Lawrence Bragg in 1912 at Cambridge,
England) are used in x-ray diffraction analysis. The mathematics of such
equations is beyond the scope of the current context.

Based on the information provided by x-ray analysis on numerous
substances, it is clear that one of the most powerful tools for understanding the structure of crystals is the concept of symmetry. Symmetry in
a crystal means that we can interchange any two parts of the crystal and
produce a replica of the original crystal. We can use this to visualize an
endless array of atoms arranged in an orderly pattern. It is the symmetry
or regularity in the spatial arrangement of atoms in a substance that give
it the notable features of a crystal. The smallest repeating unit in the
arrangement of atoms in a crystal is called the "primitive unit cell" by crys-
tallographers (Wood, 1964). A primitive unit cell is a very small structural
unit, equivalent to a building block, which is regularly repeated
throughout the crystal. These unit cells are repeated, side by side, in all
directions throughout the entire lattice of the crystal. In a "perfect crystal"
every small piece of the whole crystal is just like every other small piece.
Every place in the crystal lattice looks like every other place.

Two essential properties of a primitive unit cell are required for growth of a
crystal lattice (Wood, 1964). First, the primitive unit cell must be a particular geometrical shape which will enable it to pack together with neighboring unit cells in order to fill completely the space of the crystal lattice.
Second, it must be possible to construct mentally the whole crystal lattice
using what is called translational symmetry. Translational symmetry is a
geometrical process involving the translation of the primitive unit cell in a
direction parallel to one of its edges through a distance equal to the length
of that edge. In this way, the primitive unit cell will inhabit the location of a

neighboring cell. A clear graphical example of translational symmetry will be given later when we discuss the growth of a crystal lattice.

Regarding the first requirement, what possible shapes of a primitive unit cell will fulfill this requirement? As we examine the various possibilities, it appears that the only shapes which fit together to fill completely the space of a crystal lattice are the parallelopipeds. A parallelopiped is defined as a geometrical solid with six faces, twelve edges and eight corners, in which all opposite faces and edges are parallel. The simplest example of a parallelopiped is the cube – all the edges are the same and all angles are equal to 90°. Other types of parallelopipeds are constructed by varying the lengths of the edges and/or changing the degrees of the internal angles. In all, there are seven general classes of parallelopipeds (refer to Figure 6 for details) used as primitive unit cells. These seven classes are commonly referred to as different crystal systems which can be further divided into thirty-two point groups. Each point group represents a unique element of symmetry in the crystal. (Note: there is an inconsistency among authors as to the exact number of different crystal systems. For simplicity, we will use seven systems for the current discussion.)

Although other geometrical shapes such as the triangular prism or the hexagonal prism are able to pack together to fill the space of a crystal lattice completely, they are not considered to be primitive unit cells because they do not fulfill the second requirement of translational symmetry. For example, it is not possible to construct a lattice of triangular prisms from one triangular prism translating itself in a direction parallel to one of its edges. However, if a pair of triangular prisms (making a parallelopiped) is used as the primitive unit cell, translational symmetry becomes possible. A similar situation exists for hexagonal prisms. As stated earlier, it turns out that, based on the two requirements for a primitive unit cell, the only possible shapes which can be used are the parallelopipeds.

One way to comprehend the various types of parallelopipeds is to understand how to construct them from the cube, the simplest parallelopiped. In each parallelopiped there are three lengths of edges which can change. These are displayed in Figure 6. The lengths of the three edges are designated as "a," "b" and "c" and the angles between each pair of edges are represented by α, β and γ (Greek letters called "alpha," "beta" and "gamma"). For the cube $a = b = c$ and $\alpha = \beta = \gamma = 90°$. To construct other parallelopipeds, we can either vary the lengths of edges and/or change the degrees of the angles between the edges (refer to Figure 6).

Figure 6. The Seven Classes of
Parallelopipeds Used as Primitive Unit Cells

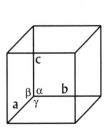
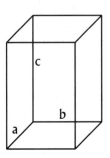
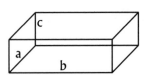

Cubic Cell	Tetragonal Cell	Orthorhombic Cell
a = b = c	a = b ≠ c	a ≠ b ≠ c
α = β = γ = 90°	α = β = γ = 90°	α = β = γ = 90°

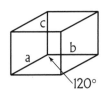
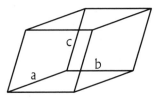

Hexagonal Cell	Rhombohedral Cell
a = b ≠ c	a = b = c
α = β = 90°, γ = 120°	α = β = γ ≠ 90°

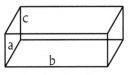
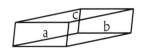

Monoclinic Cell	Triclinic Cell
a ≠ b ≠ c	a ≠ b ≠ c
α = γ = 90°, β > 90°	α ≠ β ≠ γ

Adapted from Wood, E.A. (1964). *Crystals and light: An introduction to optical crystallography.* New Jersey: D. Van Nostrand Company, Inc., p. 14

Generation of a Crystal Lattice from a Primitive Unit Cell

The repetitive array of primitive unit cells in a crystal is called a crystal lattice, and intersections of the lattice lines are called lattice points. By definition, a primitive unit cell has eight corners and at each of these corners a lattice point is situated corresponding to the location of an atom, molecule or ion. In a lattice structure each lattice point is shared by eight primitive unit cells. Thus 1/8 of each lattice point lies within a given unit cell. Since there are eight corners to a primitive unit cell, there are eight of these 1/8 pieces, or a total of one net lattice point. Sometimes however a unit cell is chosen which is double (or more than double) the size of a primitive unit cell. These larger unit cells are often used because they also reflect the symmetry of the crystal lattice. Two common examples are the body-centered and face-centered unit cells (Zumdahl, 1989). In the body-centered structure one atom is in the center of the unit cell in addition to all the eight corners. In the face-centered unit cell (also called the "closest packing" of spheres) one lattice point is situated at each corner and one lattice point is situated at the center of each face. The lattice points at the center of each face are shared by two unit cells, thus each unit cell has 1/2 of the lattice point. However, it should always be remembered that both the body-centered and face-centered unit cells can be constructed from a smaller primitive unit cell which has only the eight corner lattice points and no internal lattice points. When the primitive, body-centered and face-centered unit cells are considered, there are fourteen possible types of crystal lattices. The proof that there are only fourteen parallelopipedal crystal lattices was given by Auguste Bravais in 1848 and are called Bravais lattices in honor of his name. Seven of these are the primitive unit cells mentioned earlier. The remaining seven incorporate the structure of the body-centered and face-centered unit cells (Wood, 1964).

Repeating a primitive unit cell in all three directions to form an expanded structure is the mechanism for generation of a crystal lattice. A crystal lattice is a three-dimensional system of points, where each point represents an atom, molecule or ion. This lattice goes on for millions of cells in a crystal. The centers of the atoms, molecules or ions reside at the intersection of the lattice lines. It is important to note that crystals do not spring into existence; they grow. This growth depends upon a sufficient supply of material from the environment. When supplied with the required substances, a crystal will start to grow simultaneously in many different places and in all directions. The formation or growth of crystals can occur from any of the states of matter. For example, snow flakes grow directly from moist air which is a gas. Probably however the most familiar examples of crystal growth result from the liquid phase cooling to crystalline solid (Holden and Morrison, 1982).

From a primitive unit cell the first stage in the growth of crystal lattice will generate eight new unit cells at each of the eight original corners. This elaboration from the original eight lattice points of the primitive unit cell will create a total of sixty-four lattice points. Each lattice point on the original primitive unit cell is now a junction for eight unit cells. The addition of new unit cells to the lattice is accomplished by using translational symmetry. As described earlier, translational symmetry involves moving a primitive unit cell in a direction parallel to one of its edges. The distance moved is equal to the length of its edge. Figure 7 gives an example of crystal growth using a primitive unit lattice with all edges the same and all angles equal to 90°. This same pattern of crystal growth applies to all seven classes of parallelopipeds.

Growth of a crystal lattice will continue in all directions simultaneously unless obstructed or blocked. Usually the growth of a crystal is perfect in its early stages, but as growth continues most crystals develop flaws which are scattered throughout its structure, spoiling their otherwise unending perfection. One of the most obvious ways in which crystal growth is distorted is by presence of impurities in the growing medium of the crystal. Other ways include growth of the crystal lattice in one direction more rapidly than in other directions and interference by external forces.

The Constitution of the Universe is Reflected in the Structure of a Crystal Lattice

We have now arrived at a fundamental understanding of both the internal structure of crystals and the growth of a crystal lattice. Using these principles, we can draw a complete correspondence with the numerical pattern of the Constitution of the Universe. The first unit of the Constitution of the Universe is represented in Rig Veda by the syllable "A-K." According to the Vedic Literature, "A-K" represents a state of fullness or all possibilities, from which the entire universe is created. Similarly, crystal growth begins with a

Figure 7. Locating the Numerical Structure of the Constitution of the Universe in the Primitive Unit Cell and Growth of a Crystal Lattice

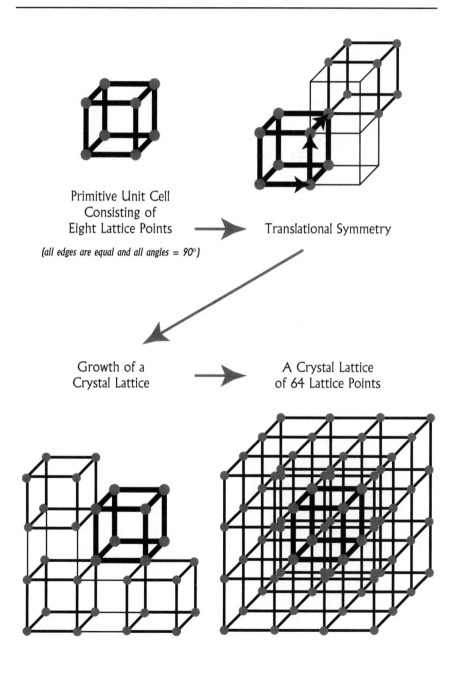

Primitive Unit Cell
Consisting of
Eight Lattice Points ⟶ Translational Symmetry

(all edges are equal and all angles = 90°)

Growth of a
Crystal Lattice ⟶ A Crystal Lattice
of 64 Lattice Points

single atom, molecule or ion which constitutes the basic building block of the primitive unit cell and crystal lattice.

The next level in the Constitution of the Universe is the unfoldment of the first syllable "A-K" into the eight syllables or prakritis of the first pada (refer to Figure 1). Likewise, the primitive unit cell of a crystal structure arises from the orderly arrangement of atoms, molecules and ions. In all parallelopiped structures of the primitive unit cell there are eight corners or lattice points. Each lattice point in a primitive unit cell represents an individual atom, molecule or ion. We propose that the orderly arrangement of the eight lattice points, which define all primitive unit cells, corresponds to the sequence of eight prakritis of the first pada of Rid Veda. From a primitive unit cell all other types of unit cell are constructed (e.g., the body-centered and face-centered structures). These in turn define crystal systems, symmetry groups and Bravais lattices.

From the eight syllables of the first pada of Rig Veda, the next level of expansion in the Constitution of the Universe is the interpretation of each of these eight syllables with respect to rishi (knower), devata (process of knowing) and chhandas (known). This produces a total of twenty-four units. Likewise the location of each lattice point in a primitive unit cell can be determined with respect to the three dimensions of space. For reasons that derive from the mathematics of crystallographic analysis, a coordinate system is set up to describe the location of lattice points. The position of any point within the coordinate system of a parallelopiped is determined according to the length of the edges and the angle between them. For more complex aspects of crystal structure the coordinate system is developed into what are called Miller indices.

The location of a lattice point with respect to the first dimension or first axis of the coordinate system represents the most abstract information. To determine the location of this lattice point we could select the length of the edge labeled "a" and the degrees of the angle α for any of the parallelopipeds in Figure 6. We propose that the location of a lattice point along this first axis matches the abstract and subtle nature of rishi. To determine the location of the lattice point with respect to the second dimension or second axis of the coordinate system we could select the edge labeled "b" and the angle β for each parallelopiped in Figure 6. Because the second dimension acts as a transition between the first and third dimensions, it corresponds with the qualities attributed by devata to connect and to act as a transitional state. To specify the location of the lattice point in three dimensions or the third axis of the coordinate system, we could select the last edge, "c," and the last angle, γ, for each primitive unit cell displayed in Figure 6. The third dimension represents the most concrete expression and is thus aptly equated with the material nature of chhandas. Together, the lengths of the edges and the angles between the edges determine the location of all eight lattice points of a primitive unit cell in three dimensions. This relates well with the interpretation of the eight prakritis of the Constitution of the Universe with respect to rishi, devata and chhandas.

In the final stage of the Constitution of the Universe the eight syllables of the first pada of Rig Veda expand into sixty-four units. A similar numerical structure is also present in the growth of a crystal lattice from a primitive unit cell. As described earlier, the first level in the growth of a crystal lattice is the generation of a new unit cell at each of the eight corners of the original primitive unit cell. A crystal lattice at this stage in its growth contains a total of sixty-four lattice points corresponding precisely with the elaboration of the eight prakritis to sixty-four syllables in the Constitution of the Universe. Further, the location of these sixty-four lattice points can be determined with respect to the three dimensions of space, thus comprising a total of 192 units. Similarly, in the final level of the Constitution of the Universe each of the sixty-four syllables of Rig Veda are interpreted with respect to rishi, devata and chhandas (a total of 192 units).

In summary, we have presented an elementary understanding of crystal structure and growth of a crystal lattice. Further, we have demonstrated that the basic structure of early crystal growth reflects the numerical structure of the Constitution of the Universe. The essence of a crystal is its ordered structure and symmetry which reflects the same order of the primal mechanics of creation. Based on these principles, crystals grow to form the myriad shapes and sizes of virtually all solids of our planet and universe.

CHAPTER 4:

BIOLOGY AND THE
CONSTITUTION OF THE UNIVERSE

SECTION I

DNA AND THE GENETIC CODE

Amid the immense diversity of biological life found throughout the evolutionary chain, a fundamental unifying entity exists called deoxyribonucleic acid, commonly referred to as DNA. DNA is nature's universal language of telling a seed or egg how to be a rose, a spider, a dolphin, a bacterium or a human. Amazingly, the same basic structure of DNA is prevalent for all species and has undergone very few modifications throughout biological evolution. The intelligence contained in the DNA, referred to as the genetic code, has been described as the blueprint or masterplan, containing the information for growth, reproduction, metabolism and development of all organisms. In this section we will provide an understanding of the basic principles of DNA and demonstrate that the structure of the genetic code exactly matches the numerical template of the Constitution of the Universe.

Discovery of DNA

Historically it was not until the end of the nineteenth century that biologists knew heredity was associated with the particular structure in the nucleus of cells known as a chromosome. Each chromosome is composed of two general types of material: proteins and DNA. However, biologists did not know which of these two large molecules was responsible for the transmission of genetic information from one cell to another cell or from one generation to the next generation. Prior to 1953 it was generally believed that only proteins had the ability to house genetic information. In the early 1950s the American geneticist, James Watson, went to England to work with Francis Crick of the Cavendish Laboratory of Cambridge to decipher the molecular structure of DNA. Building on the research of other laboratories, they were able to unravel the molecular structure of DNA and provide generally accepted evidence that it held the answers to the secrets of heredity. This accomplishment is ranked as one of the most significant breakthroughs in the history of biology. Since that time, DNA research has become the central focus of basically all areas of biological science.

As already mentioned, DNA along with associated proteins, comprise what are known as chromosomes. All organisms have a unique number of chromosomes which often occur in pairs. For humans we have a total of forty-six chromosomes or twenty-three pairs in every cell of our body. One member of each pair of chromosomes is of maternal origin and the other member is of paternal origin. In each chromosome is a molecule of DNA which contains what are called genes. Typically there are several thousand genes per chromosome. A gene is simply a particular segment of the DNA containing the information to synthesize various compounds and substances. Like a recipe, they tell the body how to grow and what to make. Every trait (e.g., eye color, height, etc.) and biochemical process in the body is controlled by genes. They are the basic unit of heredity.

Four Nucleotide Bases are the Alphabet of DNA

Structurally DNA is composed of a number of subunits which combine together linearly to form a long thread-like chain. The whole molecule of DNA consists of two such chains coiling around each other as in a spiral staircase or double-helix. Each subunit in DNA, a deoxyribonucleotide unit, is composed of three parts: a nitrogenous base, a sugar and one or more phosphate groups. The name of the sugar is deoxyribose. Linked to the deoxyribose sugars are phosphate groups and these together form the backbone structure of DNA. This backbone structure of DNA remains invariant throughout the whole molecule. The variable part of DNA is found in the nitrogenous bases which are derivatives of either of two molecules, called purines and pyrimidines. There are four nitrogenous bases in the DNA: adenine (A), guanine (G), thymine (T) and cytosine (C). The letter in parenthesis after each name is the symbol which is commonly used to denote that compound. Adenine and guanine are from the purine class of molecules and thymine and cytosine from the pyrimidine class.

One of the key aspects of DNA is the fact that the four nucleotides give DNA a variable nature. Each subunit of the DNA can have any one of these four nucleotides. Another critical aspect is how the two long chains of DNA coil around each other to form the double-helix shape. The chains run in opposite directions and are held together through chemical interactions of the nitrogenous bases forming the cross-rungs of the double helix. This chemical interaction is referred to as "base-pairing." There are only certain ways or "rules" in which the four bases can interact with one another. Adenine will always interact or pair with thymine (symbolized by A—T), and guanine will always pair with cytosine (symbolized by G—C). Together these are the two types of base-pairs which occur between the two strands of the DNA double-helix. Hence, although the sequence of bases on the one strand of the DNA is not identical to the other strand, the two strands are complementary. By knowing the sequence of bases on one strand, we can calculate the sequence of bases of the other strand by base-pairing rules. This means that in DNA the total number of thymines equals the total number of adenines and the total number of guanines equals the total number of cytosines. The DNA backbone (composed of a deoxyribose sugar and one or more phosphates) is located on the outside of the helical structure.

The precise sequence of the four bases in the DNA carries the genetic information. This information is used to specify how to create proteins. Each protein is composed of a string of amino acids joined together. The arrangement or sequence of these amino acids is specified by a corresponding sequence of the nitrogenous bases in DNA. There are, however, twenty naturally occurring amino acids and only four different bases. This means that there is a shortage in bases to code for all twenty amino acids. Thus, to obtain enough different combinations of bases to code for all amino acids we need to form sets of bases. A set of two bases will give a total of sixteen different variables (mathematically, $4^2 = 16$). However, this is still not enough. A set of three bases will give sixty-four different variables (mathematically, $4^3 = 64$) which adequately encodes for all amino acids.

These sixty-four sets of three bases each are called codons. Each codon specifies only one amino acid, but, because there are more codons than there are amino acids, the code is degenerate. Thus many amino acids are designated by more than one codon. Understanding the sequence of codons in the DNA would give the sequence of amino acids constituting a protein; proteins in turn provide the building blocks of the body. The knowledge of what codon(s) specify a particular amino acid is referred to as the genetic code. A summary of the basic concepts of DNA is presented in Figure 8.

One way to comprehend the DNA is to think of the human body as a book. The alphabet of the book consists of four letters, A, C, G and T, corresponding to adenine, cytosine, guanine and thymine, respectively. These letters combine together to form words, which correspond to codons in the DNA. Codons (words) combine together in a logical order to form sentences, corresponding to proteins. The sentences form the chapters of the book corresponding to the various organs and organ systems of the entire physiology.

Transcription and Translation are Responsible for the Expression of Genetic Information

To understand how the information contained in the genetic code is used to create proteins, we need to introduce a few new chemicals. It is important to realize that the actual process of protein synthesis occurs by the coordinated effort of over one hundred molecules and enzymes. For a basic understanding, however, we need only to be familiar with ribonucleic acid, generally known as RNA. Although RNA is similar to DNA, there are three main differences. First, the sugar molecule in RNA is called ribose instead of deoxyribose as in DNA. Second, the nitrogenous base thymine is replaced by another base called uracil (U). Hence, the base-pair adenine-thymine (A—T) in DNA changes to adenine-uracil (A—U) in RNA. Chemically the only difference between thymine and uracil is that thymine possesses a methyl group which uracil lacks. Third, RNA is a single helical molecule composed of a single strand whereas DNA is a double-helix molecule composed of two strands coiled around each other.

There are several types of RNA molecules. One of these, messenger RNA (mRNA), takes the information from a specific gene on the DNA (located in the nucleus of the cell) to another part of the cell where the information is used to specify which amino acids are to be joined together to create a

Figure 8. The Eight Codon-Codon Base-Pairs
of the DNA Double Helix and the Sixty-Four
Codons of the Single-Stranded DNA

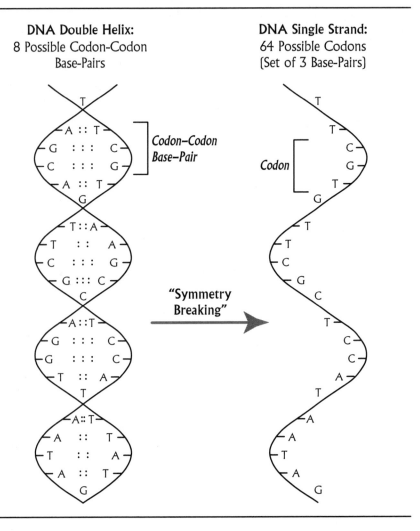

DNA Double Helix:
8 Possible Codon-Codon
Base-Pairs

DNA Single Strand:
64 Possible Codons
(Set of 3 Base-Pairs)

"Symmetry
Breaking"

**Genetic Code has
4 Letters (Base-Pairs):**

- Adenine (A)
- Thymine (T)
- Guanine (G)
- Cytosine (C)

Base-Pairing Rules:

- Adenine always
  pairs with Thymine
- Guanine always
  pairs with Cytosine

protein. For this to happen, a sequence of biochemical events occur which are generally divided into two categories: transcription and translation.

As the name implies, "transcription" is the process by which genetic information is transferred from a DNA template to an mRNA molecule. One way to view this process is to think of DNA as a storehouse of information and transcription as a photocopier which makes a copy of a portion of this information. This copy is what is called mRNA. The biochemical mechanisms by which this occurs are as follows. First, a segment of the DNA molecule "unwinds" itself through the actions of a constellation of enzymes. This segment of the DNA corresponds to a particular gene containing the information for the specific protein which needs to be synthesized. The exposed region of the DNA contains a sequence of bases which serve as a template for assembling an mRNA molecule. This is accomplished through a process called complementary base-pairing. As already described, the rule of base-pairing is that adenine always pairs with thymine (or with uracil in RNA) and guanine with cytosine. Hence, given the order of bases on one strand of the exposed DNA, a "complementary" order of bases on an RNA molecule can be assembled. Specific sections on the DNA guide the enzymes where to start assembling the mRNA molecules and where to stop. Figure 9 summarizes a simple version of DNA transcription and translation.

Once the mRNA molecule has been assembled, it departs from the DNA molecule and travels out of the nucleus of the cell to another part of the cell where a ribosome is located. There are many ribosomes in each cell and their primary function is to synthesize proteins according to the instructions carried on the mRNA molecule. The mechanism by which the genetic information in mRNA is used to create proteins is called "translation." Another way to understand "translation" is to think of it as a process involving the transfer of information from one language (the four-lettered alphabet of bases in the mRNA) into another language (amino acids which are the building blocks of proteins).

At the ribosome the mRNA is combined with a series of small molecules called transfer RNA's (tRNA's). The tRNA molecule acts as "translator" between the language of base sequences in an mRNA molecule and the language of amino acids making up the structure of proteins. For the sake of simplicity, we can analyze each tRNA molecule as having two important parts. First, each tRNA molecule contains a sequence of three bases (called an anticodon) that interact with a codon on the mRNA according to

Figure 9. Mechanism of Protein Synthesis: DNA Transcription and Translation

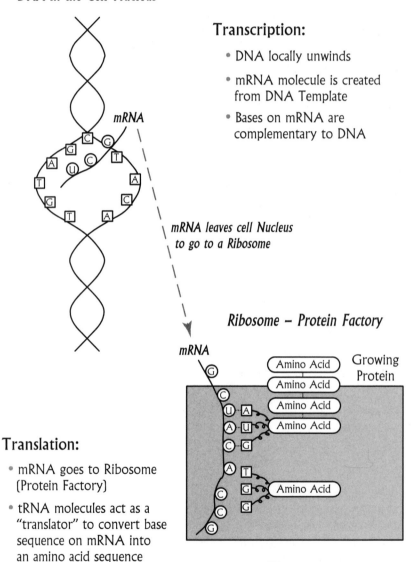

DNA in the Cell Nucleus

Transcription:

- DNA locally unwinds
- mRNA molecule is created from DNA Template
- Bases on mRNA are complementary to DNA

mRNA

mRNA leaves cell Nucleus to go to a Ribosome

Ribosome – Protein Factory

mRNA

Amino Acid
Amino Acid
Amino Acid
Amino Acid

Growing Protein

Amino Acid

Translation:

- mRNA goes to Ribosome (Protein Factory)
- tRNA molecules act as a "translator" to convert base sequence on mRNA into an amino acid sequence
- a sequence of amino acids is called a protein

tRNA   Amino Acid

complementary base-pairing rules. This means that for each codon on mRNA there is a matching tRNA that carries an anticodon having bases which are complementary to the bases on the mRNA codon. Second, each tRNA molecule has a place where one amino acid can be attached.

Key to understanding "translation" is the relationship between the anticodon and the attached amino acid on a tRNA (refer to Figure 9). Binding of the tRNA anticodon to the mRNA codon is governed by rules of base-pairing. This interaction is independent of the amino acid that is attached to the tRNA. As a result, whatever amino acid is attached to the tRNA during protein synthesis will be incorporated into the growing protein chain. The fidelity of protein synthesis is thus, in part, dependent on the proper attachment of each amino acid to the correct tRNA. Specific enzymes and structural features of each tRNA are responsible for binding the correct amino acid that will correspond to the proper mRNA codon as determined by the genetic code. Hence, given a sequence of codons on an mRNA, the ribosome is capable, via tRNA, to assemble a corresponding sequence of amino acids which eventually become a protein. In the mRNA codon sequence there are specific codons that signal the ribosome to start and to stop protein synthesis (out of the sixty-four codons in the genetic code sixty-one encode amino acids and three are termination codons). Once the synthesis of a protein is completed, the protein can be used as a building block according to the shortage or surplus of compounds needed by the cell.

The Genetic Code of DNA is Identical to the Constitution of the Universe

We have described some of the basic principles of the DNA and explained how the genetic code contains all the information for growth, reproduction, maintenance and development of every living organism. Using these fundamentals, we will now proceed to locate in the genetic code of the DNA the numbers eight, sixty-four and three of the Constitution of the Universe. Our first connection relates to the initial level of unfoldment in the Constitution of the Universe, from the first syllable "A-K" into the eight syllables of the first pada (refer to Figure 1). We propose that the eight unique codon-codon base-pairs in the DNA double helix correspond to the eight prakritis of the Constitution of the Universe. The eight different combinations of codon-codon base-pairs are summarized in Table 7.

Table 7. Locating the Fundamental Structure of the Constitution of the Universe in the Genetic Code of DNA

Rishi Interpretation – DNA:

The Eight Codon–Codon Base–Pairs:[a]

BBB	BBD	BDB	BDD	DBB	DBD	DDB	DDD

The Sixty–Four DNA Codons:[b]

AAA	AAG	AGA	AGG	GAA	GAG	GGA	GGG
AAT	AAC	ACA	AGC	GAT	GTG	GGT	GGC
ATA	ATG	AGT	ACG	GTA	GAC	GCA	GCG
TAA	ATC	ACT	ACC	GTT	GTC	GCT	GCC
ATT	TAG	TGA	TGG	CAA	CAG	CGA	CGG
TTA	TAC	TCA	TGC	CAT	CTG	CGT	CGC
TAT	TTG	TGT	TCG	CTA	CAC	CCA	CCG
TTT	TTC	TCT	TCC	CCT	CTC	CCT	CCC

Devata Interpretation – mRNA:

The Eight Types of mRNA Codons:

EEE	EED	EDE	EDD	DEE	DED	DDE	DDD

The Sixty–Four mRNA Codons:

UUU	UUC	UCU	UCC	CUU	CUC	CCU	CCC
UUA	UUG	UGU	UCG	CUA	CAC	CCA	CCG
UAU	UAC	UCA	UGC	CAU	CUG	CGU	CGC
AUU	UAG	UGA	UGG	CAA	CAG	CGA	CGG
UAA	AUC	ACU	ACC	GUU	GUC	GCU	GCC
AAU	AUG	AGU	ACG	GUA	GAC	GCA	GCG
AUA	AAC	ACA	AGC	GAU	GUG	GGU	GGC
AAA	AAG	AGA	AGG	GGA	GAG	GGA	GGG

Chhandas Interpretation – Amino Acids:

The Eight Types of tRNA Codons:

EEE	EED	EDE	EDD	DEE	DED	DDE	DDD

The Sixty–Four Amino Acids Corresponding to Each mRNA Codon:

phenylalaine	phenylalaine	serine	serine	leucine	leucine	proline	proline
leucine	leucine	cysteine	serine	leucine	histidine	proline	proline
tyrosine	tyrosine	serine	cysteine	histidine	leucine	arginine	arginine
isoleucine	stop	serine	tryptophan	glutamine	glutamine	arginine	arginine
stop	isoleucine	threonine	threonine	valine	valine	alanine	alanine
asparagine	methionine	serine	threonine	valine	aspartic acid	alanine	alanine
isoleucine	asparagine	threonine	serine	aspartic acid	valine	glycine	glycine
lysine	lysine	arginine	arginine	glycine	glutamic acid	glycine	glycine

(Note: mRNA codon AUG is usually the start signal in addition to coding for methionine.)

[a] B = Adenine-Thymine base pair; E = Adenine-Uracil base pair; D = Cytosine-Guanine base pair.
[b] A = Adenine; T = Thymine; U = Uracil; G = Guanine; C = Cytosine.

These eight unique codon-codon base-pairs are calculated as follows. As explained earlier, a "codon" is a sequence of three bases, and a "base-pair" is the way in which the four bases interact (i.e., adenine always pairs with thymine [A—T] and guanine always pairs with cytosine [G—C]). Hence there are only two types of base-pairs. Each base-pair can be viewed as a single unit where there is no distinction between adenine and thymine or cytosine and guanine. In this way the total number of different combinations in a codon composed of a sequence of three *base-pairs* turns out to be exactly eight (mathematically, $2^3 = 8$). The eight different codon-codon base-pairs are associated with the **double**-stranded DNA, in which base-pairs form the "cross-rungs" of the structure.

From the eight basic syllables of the first pada of Rig Veda, the next level in the expansion of the Constitution of the Universe is the interpretation of each of these eight syllables with respect to rishi (knower), devata (process of knowing) and chhandas (known). This produces a total of twenty-four units. Likewise there are three basic aspects in DNA expression which correspond to the interpretations of rishi, devata and chhandas (Wallace, 1986 and Wallace, 1993). These three basic aspects are commonly referred to as the "DNA dogma" DNA → mRNA → protein. The first of these three aspects is the DNA itself, the abstract information which remains in the cell nucleus and is not directly involved in protein synthesis. This matches the nature of rishi which embodies the qualities of abstractness, intelligence and non-involvement with activity. The second aspect is mRNA in dynamic activity of transcription and transport of the information inherent in DNA from the cell nucleus to the ribosomes of the cell. This corresponds very well with the attributes of the devata interpretation which are dynamism and action. The last aspect is the actual amino acids of the synthesized protein corresponding to the anti-codons of the tRNA, which result from the process of translation. This is equated with the expressed or material nature of the chhandas interpretation. These three aspects, when viewed with respect to the eight codon-codon base-pairs, comprise a total of twenty-four units, corresponding to the twenty-four units of the Constitution of the Universe.

In the next stage of elaboration of the Constitution of the Universe, the eight syllables of the first pada of Rig Veda expand into sixty-four units. Similarly the genetic code of DNA also contains exactly sixty-four codons. As described earlier, the reasoning behind Nature's selection of sixty-four codons is based on the need to code for all twenty amino acids. These sixty-four codons are associated with a **single**-strand of DNA, in which each codon is made up of a sequence of three bases.

The final correspondence with the Constitution of the Universe stems from the sixty-four syllables of Rig Veda being interpreted with respect to rishi, devata and chhandas, thereby making a total of 192 units. Likewise we can view each of the sixty-four codons of the genetic code with respect to how they are expressed in the DNA, in the RNA and in amino acids (a total of 192 units).

At this stage, it is interesting to draw a parallel with the concept of "symmetry breaking" in physics. According to physics, the principle of symmetry breaking locates hidden symmetries in nature. In quantum physics distinct forces and particles on one level become fundamentally indistinguishable when viewed at a deeper unified level. For example, the unified electroweak theory unifies the weak and electromagnetic forces. At this unified level, the electron and neutrino, the up-quark and the down-quark, etc. are believed to be indistinguishable. The apparent difference between the weak and electromagnetic forces is thus a result of spontaneous symmetry breaking. Likewise a similar situation appears to be at work in the DNA. In the double-helix of DNA the four bases exist in base-pairs with no distinction between adenine and thymine or between cytosine and guanine. It is not until the expanded stage, "symmetry breaking," of the single-stranded DNA that the four bases are seen as distinct and thereby creating the sixty-four different codons.

In summary, we have located in the structure of the genetic code of DNA a pattern which is exactly the same as the numerical template of the Constitution of the Universe. This is a particularly exciting discovery because it makes it easier to conceptualize how the Constitution of the Universe is actually located in everyone's physiology. The correspondence between the Constitution of the Universe and the numerical structure of the genetic code of DNA was recognized by the first author of this book and was presented to the faculty and students at Maharishi International University, Fairfield, Iowa on April 8th 1992. Shortly after this discovery other independent researchers arrived at similar conclusions. These individuals include Dr. G. Brown, Dr. R. Chalmers, Dr. T. Nader (Nader, 1995) and Dr. H. Sharma (Sharma, 1993). Prior to the writing of this book, this correspondence along with the superstring theory of quantum physics (refer to Chapter 2, Section I) were the only two recognized parallels between modern science and the numerical structure of the Constitution of the Universe.

The Structure of the Genetic Code Matches the I Ching

It is also worth noting that a number of independent investigators have published evidence that the mathematical structure of the DNA molecule is strictly analogous to the numerical structure of the most revered text of ancient Chinese wisdom, the I Ching. One of the first individuals to make this association between the I Ching and the DNA was a young post-doc in molecular biology named Harvey Bialy. Dr. Bialy's work, published in a small literary magazine, was expounded upon later by Gunther Stent (Stent, 1969). In addition to Bialy-Stent's work, two other individuals independently observed this same striking similarity between the I Ching and the DNA: first, Dr. Schonberger in 1973 (translated into English, Schonberger, 1992), and second, Johnson Yan (Yan, 1991). This established link between the DNA and the I Ching not only gives further credence to the universal importance of the structure of the genetic code, but also naturally suggests that there must be a strong similarity between the I Ching and the Constitution of the Universe (refer to Chapter 8, Section IV for details).

SECTION II

EMBRYONIC DEVELOPMENT

All living organisms have a more or less limited life span. Thus, for a group of organisms to survive, there must be a mechanism for the production of new individuals, to ensure the continuation of the species. This process called reproduction marks the origin or beginning of a new organism. The study of the origin and development of a growing organism is called embryology. In this section we will discuss the basic principles of embryology, which apply to most organisms, however we will limit our focus to human embryology. Further, we will demonstrate that the early stages of embryonic development reflect a structural pattern which is the same as the numerical framework of the Constitution of the Universe.

The Cellular Basis of Reproduction

The purpose of reproduction is to create new individuals that are "copies" or "replicas" of their parents. Depending on the species, these new individuals may be produced in one of two ways. In the first way, asexual reproduction, the new individuals produced are exact copies of their parents. Many organisms, such as single-celled bacteria, reproduce in this fashion. In the second way, sexual reproduction, the new individuals created are *not* exact replicas of their parents. This is the method of reproduction used by humans.

To understand why some individuals are (or are not) exact "copies" of their parents, we need to comprehend the molecular or cellular mechanisms behind such events. In all biological life, the cell is considered to be the smallest unit and building block of life. Within each cell a particular compound, called DNA (refer to previous section for details), contains a complete set of hereditary instructions for the growth, reproduction, and metabolism of the organism. Together with some specialized proteins, DNA combines to form what is known as a chromosome. All organisms have a unique number of chromosomes. In humans there are forty-six chromosomes in each somatic cell (all cells except reproductive cells). Of these forty-six, two are called sex chromosomes and the remaining forty-four are called autosomes. The sex chromosomes determine the gender of the individual (two X chromosomes in women, and one X and one Y in men). The other forty-four chromosomes (autosomes) contain genetic information responsible for the rest of the physiology. Together these forty-six chromosomes in humans exist as twenty-three pairs. The chromosomes in each pair, known as "homologous chromosomes," are similar in length and shape (except the two sex chromosomes in males – X and Y). One member of each pair of chromosomes is derived from paternal origin and the other member from maternal origin. All chromosomes are made up of two identical chromatin threads called sister chromatids.

To understand asexual and sexual reproduction we need to know how the chromosomes of the parents are passed on to the offspring. The passage of chromosomes from parent to offspring occurs through cell division. When any cell divides, it signifies the bridge to the next generation. In this process, one cell (called the parent cell) is split into two smaller cells (called daughter cells). For single-celled organisms cell division represents the process of reproduction by which new individuals are produced. However, in multicellular organisms the process of cell division could lead either to the creation of reproductive cells or to physical growth and maintenance of the individual, depending on the type of cell.

Cell division accounts for one of three major periods in the life cycle of any cell. The other two periods cover cell growth and chromosomal duplication. In cell growth a cell manufactures proteins and other needed substances. This is also a time when the cell can remain dormant. The second period in the life cycle of a cell is the duplication of chromosomes, which assures that when cell division occurs each daughter cell will contain the same genetic information. Together these three periods allow for successive generations of cells.

Mitosis and Meiosis are the Two Types of Cell Division

There are two major types of cell division which determine whether reproduction is asexual or sexual. The first type of cell division, called mitosis, produces two daughter cells which are genetically identical to one another. When cell division occurs by mitosis, the two sister chromatids of each chromosome separate. One chromatid goes to one daughter cell and the other chromatid goes to the other daughter cell. In this fashion both daughter cells receive the same forty-six chromosomes (each chromosome will contain only one chromatid). Following cell division each daughter cell will synthesize or replicate a matching chromatid for each chromosome – in preparation for another cell division. The hallmark of mitosis is the fact that the number of chromosomes remains the same from one cell generation to the next. This is the molecular basis for under-standing asexual reproduction as well as growth of multicelled organisms.

The second type of cell division, known as meiosis, produces two daughter cells which are not identical to one another. In humans the mechanism of meiosis is used only to produce a special group of cells called "germ" or reproductive cells (also known as gametes). These unique cells are produced by the reproductive organs – in humans these are the ovaries in females and the testes in males. The process of meiosis, or the formation of gametes, involves two stages. In the first stage the chromo-somes from each homologous pair separate into each of two daughter cells (twenty-three chromosomes in each). Then, in the second stage these daughter cells in turn undergo cell division. During this second cell division the sister chromatids of each chromosome separate from one another (similar to the process of mitosis). The net result of meiosis is two cell divisions, which create a total of four new cells (gametes) that each have twenty-three unduplicated chromosomes. Cells (before meiosis) having forty-six chromosomes (twenty-three pairs of homologous chro-mosomes) are known as diploid cells. Cells (after meiosis) which have only twenty-three chromosomes are known as haploid (half the genetic information). The hallmark of meiosis is that the number of chromosomes are halved from one cell generation to the next and that the gametes produced contain only half the genetic information. Restoration of the total number of chromosomes and complete genetic information occurs when two gametes unite (one from the mother and one from the father). This is the molecular basis for understanding genetic diversity in the offspring produced by humans and other sexually reproducing organisms. It also explains why we have some characteristics derived from our father and others which are derived from our mother.

A Zygote is Created by the Fusion of Male and Female Germ Cells

As mentioned earlier, the production of germ cells in humans is carried out by the reproductive organs. In females the germ cells formed by the ovaries are called ova (plural for ovum). Mature ova, known as oocytes, are produced in cycles lasting about 28 days. Compared with somatic cells, oocytes are large, non-motile and exhibit a huge cytoplasmic to nuclear ratio. In males the germ cells formed by the testes are called spermatozoa (plural for spermatozoon) or sperm. Production of mature sperm from primary spermatocytes in humans requires about sixty-four days (Spence and Mason, 1992). Mature sperm are among the smallest cells in a human body, are highly motile, and are shaped like a comet with a long tail and a head containing the nucleus. Both sperm and ova are haploid cells, which means that they each contain only half the information necessary for the development of a new individual. Only when a sperm and ovum unite during sexual intercourse do we have the complete genetic information needed for development of a new person.

During coitus (sexual intercourse) sperm from the male is deposited into the female vagina. Once the sperm has entered the female reproductive tract, it undergoes a process called capacitation. During this process the glycoprotein coat and various proteins are removed from the head of the sperm which enables it to unite with the ovum produced by the female. Capacitated sperm then move up the vagina into the uterus and reach the upper uterine tubes where an ovum is released during ovulation. If the conditions are right, the actively moving sperm will encounter a mature ovum and fertilization will take place (Moore and Persaud, 1993).

In brief, there are three main stages involved in the process of fertilization. The first stage, known as "penetration," occurs when a sperm adheres or binds to one of the specific receptor sites on the zona pellucida which surrounds the ovum. A sperm can bind to the ovum at any location. In the second stage, called "activation," the sperm undergoes an acrosomal reaction. During this reaction enzymes such as proteases and hyaluronidase are released which are needed for penetration of the zona pellucida (Spence and Mason, 1992). The final stage of fertilization, called fusion, is when a sperm actually penetrates the zona pellucida and enters into the ovum. Once a sperm has entered the ovum, a number of biochemical changes occur which prevents the entry of any additional sperm. Inside the ovum the nuclei of the sperm and ovum fuse together to create one cell. This single cell is called a zygote (from the Greek "zygotos" which means "yoked together") and marks the beginning of human development.

When a zygote is formed, it results in the restoration of the diploid number of chromosomes. Prior to fusion each germ cell was haploid (twenty-three chromosomes) and contained half the genetic information needed to create a new individual. After fusion the zygote is diploid (forty-six chromosomes) – twenty-three chromosomes from the mother and twenty-three chromosomes from the father. Since all chromosomes are present, it is possible at this stage to determine the gender of the new individual. Prior to their fusion both sperm and ova are highly specialized cells, yet after fertilization they create a zygote which is regarded as one of the most unspecialized and most undifferentiated cells.

Early Stages of Embryonic Growth

Once the zygote is created, it immediately begins to divide by mitosis. In about thirty-eight weeks (266 days) this single-celled zygote is transformed into a 200 thousand million-celled individual ready for birth. This time of massive growth accounts for a substantial amount of our developmental changes. In humans (as well as mammals, birds and reptiles) this process takes place within the body of the female.

When a zygote begins to divide, or to undergo "cleavage," there are a number of notable characteristics. Compared with normal cells, the rate of cell division within the zygote is very rapid. During the early stages of these cell divisions there is no increase in the overall mass of the zygote – the size of the individual cells is rapidly reduced as the cells continue to divide. Preceding each cell division, there is an increased synthesis of nuclear material and no increase in the amount of cytoplasm. As a result the huge cytoplasmic to nuclear ratio, which was a hallmark of the ovum, is gradually restored to a normal balance.

While the zygote is dividing, it also begins to move down the upper portion of the uterine tube to the uterus. The uterus is a thick-walled organ in which the dividing zygote will eventually reside and grow. As the zygote undergoes cleavage, it follows a binary sequence. When the zygote first divides, it creates two daughter cells, then the two daughter cells divide which gives four cells, and so on. The cleavage remains regular and distinct until the eight-cell stage (Smith and Williams, 1984). At this level all eight cells are undifferentiated and unspecialized. After this stage the cells become polyhedral in shape and fit together by a process known as compactification. This process permits the formation of what are called gap and tight junctions allowing the cells to communicate and transport substances between themselves. Typically it takes about three to four days for the zygote to divide internally into eight cells, by which time it has moved down the uterine tube and is ready to enter the uterus.

Cleavage after the eight-cell stage becomes less regular, and the growing number of cells become organized into a mulberry-shaped sphere. This mass of cells is called a morula and typically contains between thirty-two and sixty-four cells. After the morula stage, cell division continues and the cells become arranged into a fluid-filled ball called a blastocyte containing between 500 and 2000 cells. By this time two events have occurred: first, the zona pellucida on the blastocyte has disintegrated which allows it to expand in size; second, the endometrium (side or wall) of the uterus has been prepared for implantation of the blastocyte by high levels of progesterone and estrogens. The blastocyte adheres to the endometrium and releases enzymes which begin to digest a portion of the endometrical surface. This digestion allows the blastocyte to implant or burrow itself into the endometrium. Digestion of the endometrical cells provides additional fluids and nutrients which help the developing embryo to grow.

After implantation of the blastocyte in the endometrium, massive growth and development of the embryo continue. By the eighth week the fetus has recognizable features of a human – hands, feet, face and all major organs. At about thirty-eight weeks the fetus is ready for birth. Following birth the new individual continues to grow and develop as he/she progresses through infancy, childhood, puberty and adolescence.

Embryonic Development Resembles the Structure of the Constitution of the Universe

Early embryonic growth bears a close resemblance to the numerical structure of the Constitution of the Universe. Our first correspondence relates to the first syllable of Rig Veda "A-K." According to Rig Veda, "A" contains in seed form the complete knowledge of natural law, a state of all possibilities, which collapses to the point value of "K." The letter "K" represents a state of zero possibilities. Likewise in embryonic development the zygote represents the first cell ("K") of the new individual arising from the inherited pool of genetic possibilities ("A"). Each zygote is one of many possible mixtures of maternal and paternal chromosomes.

The next level in the Constitution of the Universe is the unfoldment of the first syllable "A-K" into the eight syllables or eight prakritis of the first pada (refer to Figure 1). A similar numerical pattern is also present in the early stages of embryonic growth. During cleavage the zygote transforms from a single cell into a multicellular cluster. The first cleavage division results in two daughter cells. Thereafter cleavage remains regular and distinct until the dividing zygote consists of eight cells. As described earlier, this eight-cell stage appears to have significance in human embryology. We propose that the production of these eight cells from the single-celled zygote reflects the unfoldment of the eight syllables of the first pada of Rig Veda arising from "A-K."

From these eight syllables of the first pada of Rig Veda, the next level of expansion in the Constitution of the Universe is the interpretation of each of these eight syllables with respect to rishi (knower), devata (process of knowing) and chhandas (known). This produces a total of twenty-four units. Likewise there are three basic stages to the life cycle of a cell corresponding to the interpretations of rishi, devata and chhandas. The first stage in the life cycle of a cell is called "growth." During this stage there is no substantial change in the cell except for the production of proteins and other needed substances. We propose that this stage in the life cycle of a cell matches the abstract and subtle qualities attributed to the rishi interpretation. During the second stage in the life cycle of a cell, called "synthesis," all chromosomal DNA is duplicated. This process matches the active and dynamic characteristics of the devata interpretation. The final stage in the life cycle of a cell, when the cell divides to produce two daughter cells, corresponds to chhandas which is equated with the most expressed or material properties of an entity. Observing these three stages with respect to the life cycle of each of the first eight cells produced from cleavage of the zygote, we have a total of twenty-four units, corresponding to the twenty-four units of the Constitution of the Universe.

In the next stage of elaboration of the Constitution of the Universe the eight syllables of the first pada of Rig Veda expand into sixty-four units. Similarly cleavage divisions in the zygote continue after the eight-cell stage. In the initial stages of this continued growth, the mass of cells is called a morula and contains somewhere between thirty-two and sixty-four cells. The exact number of cells varies however, because cleavage divisions past the eight-cell stage become less regular. During this stage of development the morula has moved out of the uterine tube and is beginning to prepare for implantation. Even though the exact number of cells in the morula varies, we propose that because the number of cells fluctuates between thirty-two and sixty-four it seems to be a suitable match with the elaborated sixty-four syllables stage in the Constitution of the Universe.

The final correspondence with the Constitution of the Universe stems from the sixty-four basic syllables of Rig Veda interpreted with respect to rishi, devata and chhandas, thus comprising a total of 192 units. Likewise we can observe each of the cells contained in the morula in light of the three stages constituting the life cycle of a cell.

In summary, we have discussed the basic principles of the early stages of human embryonic development which provide the basis for the production of new life. Further, we have demonstrated that the numerical sequence of events characterizing the early stages of embryonic growth match the numerical structure of the Constitution of the Universe which lies at the basis of creation. Although this correspondence is based on the early stages of embryonic growth which are relatively simple, these stages also determine what will happen through the life of the organism. Following these early stages, growth continues to occur in the individual throughout the rest of embryonic development, infancy and adolescence. Adolescence ends when the individual becomes capable of beginning the process of reproduction anew.

CHAPTER 5:

MUSIC AND THE CONSTITUTION OF THE UNIVERSE

THE FUNDAMENTAL MUSICAL VIBRATION AND ITS HARMONIC OVERTONES

There is an old adage, "As in music, so in life." Since time immemorial ancient civilizations upheld this truth and recognized that music was more than mere entertainment or recreation but was intimately associated with culture and tradition for its universal attraction, beauty and power. In ancient Greece, for example, music was believed to arise from divine origin, having the magical ability to purify the mind and body and mediate transformations in Nature (Grout, 1973). Yet, along with this aesthetic and universal appreciation for music there is also an objective understanding of the nature of musical sounds. However, since most musicians have needed to understand intuitively only how to compose and play music, the majority of our objective knowledge of music theory is attributed to the work of scientists. In this section we will summarize the basic principles of music theory and relate them to the numerical structure of the Constitution of the Universe.

Description of a Musical Tone

Music is a pleasing auditory sensation arising through the medium of sound. A musical tone is a pleasant sound, different from a noise, which is not so pleasing. Although a noise and a musical tone can intermingle and be transformed into one another, they are distinct at their extremes. Noises are characterized by irregular, possibly "random" and complex motion of sound waves that are not periodic. Musical tones on the other hand are composed of simple, regular and uniform sound waves which exert a rapid periodic motion (Helmholtz, 1954). Between these two types of sounds, the musical tone is perceived as more pleasant and is therefore considered to be the simplest element of music. Each tone is represented by a letter of the alphabet corresponding to a particular frequency of sound waves.

Each musical tone is distinguished from other tones in three fundamental ways. The first characteristic used to describe a musical tone is called the quality or timbre of a sound. This property distinguishes one musical tone from another tone of the same volume and pitch. For example, quality is the distinguishing feature we hear between a human voice and a trombone or a piccolo when all these instruments play the same note at the same volume. The second peculiarity used to describe a musical tone is called the pitch of a sound. Pitch is simply the frequency or cycles/second of the sound wave producing a given tone. The greater the pitch, the sharper the sound, and the lower the pitch, the deeper the sound. The last factor used to describe a musical tone, the force of the sound, is a direct result of the amplitude of a sound wave and is responsible for how "loud" or "soft" a particular musical tone is. Together, these three peculiarities are all that are needed to distinguish between two musical tones (Helmholtz, 1954).

A Musical Tone Contains a Spectrum of Harmonic Overtones

The structure of a musical tone can be represented as a wave pattern consisting of a fundamental vibration and a series of upper partials or harmonic overtones. The fundamental vibration (also known as the first partial) is the frequency of the musical tone being played. The sound of the fundamental vibration is the main sound we hear when that musical tone is being played. For example, the fundamental vibration of the tone C has a frequency of 514 vibrations/second. Contained within a musical tone are a number of other tones, called the harmonic overtones, which are also present along with the fundamental vibration. These harmonic overtones arise as whole number multiples of the frequency of the fundamental vibration (if this were not the case, we would hear a noise rather than a musical tone). Within the internal structure of the harmonic overtones there is a complete alphabet of all tonal relationships, musical intervals and values of consonance and dissonance of music. Essentially this means that when a musical tone is played, we are actually hearing a combination of a fundamental note and a series of barely audible overtones.

Another way to clarify the concept of the harmonic overtone series is to relate it to the phenomenon of white light. When we look at sunlight we observe just one color, but when this light is split by a prism we obtain the familiar spectrum of colors. In a similar way a musical tone consists of a series of overtones. The "spectrum" of the world of sound is the harmonic overtone series (Hindemith, 1945).

The Octave is the Most Basic Musical Interval

When we listen to the different parts of the harmonic overtone series, what we are hearing is a sequence of musical intervals or tonal relationships. A tonal relationship refers to the effect created when two tones are sounded together. Since a musical tone can be expressed as a numerical frequency, an interval is generally represented as a mathematical ratio which tells the distance in pitch between two tones.

The first and most basic musical interval in the harmonic overtones is called the octave. This interval is the most rudimentary, simplest and perfect consonant in virtually all types of music worldwide. An octave interval is defined by two musical tones: a fundamental vibration and a second note that has twice the frequency. For example, if we take the tone C to be the fundamental vibration, it would have a frequency ratio of 1:1 (representing 514 vibrations/second). The octave of C would be C_1 which would have a frequency ratio of 2:1 (representing 1,028 vibrations/second). As evident from this example, any two tones that are an octave apart are given the same letter name. That is, if we start on a given musical tone, we will end on that same tone one octave higher. This is because the musical tones of an octave exhibit a very similar sound quality. In fact, the resemblance of two tones an octave apart is so strong that, to the normal listener, they produce the effect of one sound. This special feature of the 2:1 relationship has often been termed the "basic miracle of music." Throughout nature, every fundamental tone is accompanied by its second harmonic (Jeans, 1968).

In all scales, the first and last tones are an octave apart. A musical scale (from the Latin word "scala," meaning a ladder) consists of a sequence of notes. There are generally eight notes in a scale which start from a funda-

mental vibration (typically C) and end one octave higher (typically C_1).
For the purpose of this chapter we will limit our discussion to the scale of
eight notes even though a number of other scales have existed throughout
history. Irrespective of the fact that other scales can have five (pentatonic
scale used in China), twelve, eighteen, or various other numbers of tones,
the interval between the first and last notes are, by convention, always
called an octave.

Although the origins of the octave scale of eight notes has not been traced,
it has existed at least as far back as the Greco-Roman civilization and
ancient India. The ancient Greeks recognized the significance of the
number eight, even to the extent of coining the ancient proverb "all things
are eight" (Taylor, 1972). During that same era the famous Pythagoras stan-
dardized the octave set of eight notes arithmetically, forming the basic
musical scale that is prevalent throughout all Western music today. Four
principal scales have been used in Western music. First, the Pythagorean
Scale dating back to the time of Pythagoras in 550 BC. Second, the Just or
Natural Scale attributed to the Egyptian scholar Ptolemy who lived in the
first century. The word "Just" derives from the intervals that are tuned
precisely according to the proportions derived from the harmonic overtone
series. The intervals of this eight-note scale are essentially the same as the
Pythagorean Scale, but with minor deviations. The last two scales are the
Mean-Tone Scale and the Equal Temperament or Chromatic Scale.

Intervals of the Harmonic Overtone Series

As we examine the structure of the harmonic overtone series, as described
in detail above, the first interval is the musical octave. This interval occurs
throughout the harmonic overtone series wherever there is a 2:1 ratio
(e.g., 4:2, 8:4, 16:8, etc.). After the octave, the next interval in the
harmonic overtone series, called the perfect fifth, has a frequency ratio of
3:2 and represents the tone G. Following the perfect fifth, the next
interval, the second octave 4:2 is the same as the ratio of 2:1. In this way
more and more musical intervals are constructed as we progress higher
into the harmonic overtone series. However, as we construct more and
more harmonic overtones, the intervals between the individual overtones
grow smaller and smaller. A diagrammatic representation of the first sixty-
four harmonics of the tone C are shown in Figure 10. As can be observed
from this diagram, the number of overtones increases in a binary sequence

Figure 10. The First Sixty-Four Harmonics of the Musical Tone C

(Musical tones corresponding to the ratios are according to the Just or Natural Scale)

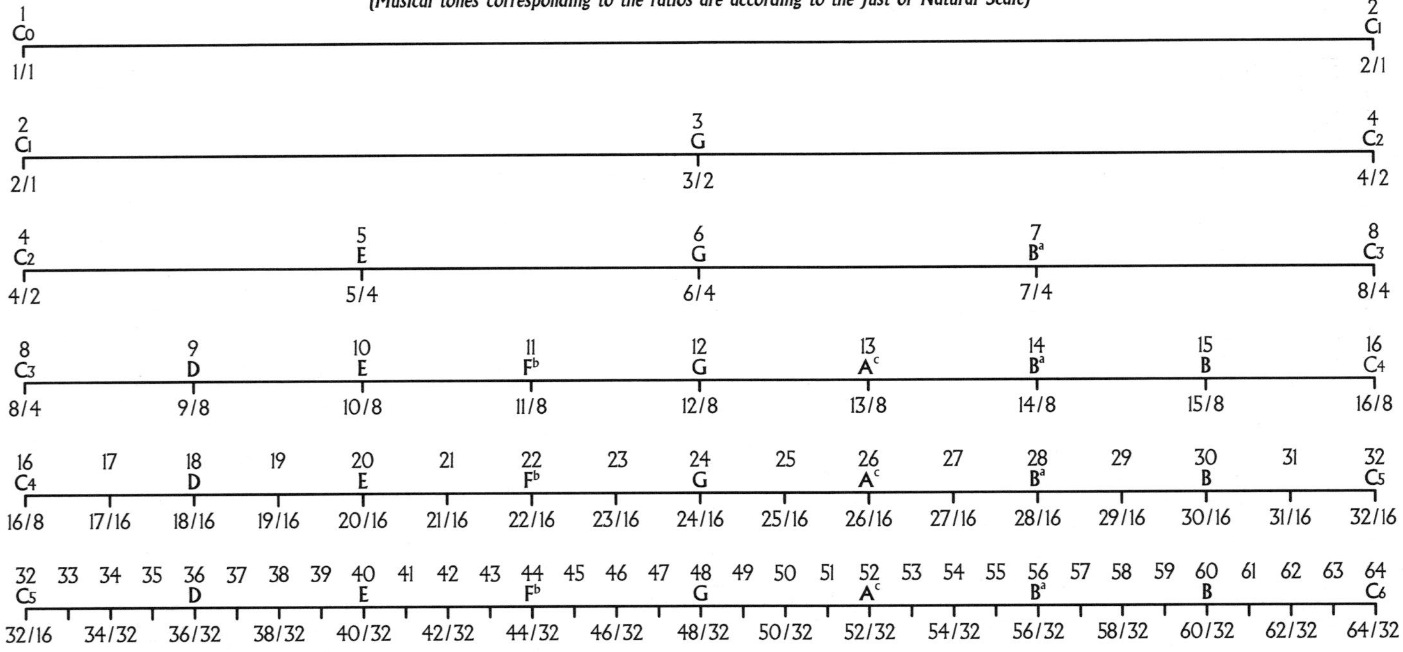

Harmonic	Tone	Ratio
1	C₀	1/1
2	C₁	2/1
3	G	3/2
4	C₂	4/2
5	E	5/4
6	G	6/4
7	B[a]	7/4
8	C₃	8/4
9	D	9/8
10	E	10/8
11	F[b]	11/8
12	G	12/8
13	A[c]	13/8
14	B[a]	14/8
15	B	15/8
16	C₄	16/8
17		17/16
18	D	18/16
19		19/16
20	E	20/16
21		21/16
22	F[b]	22/16
23		23/16
24	G	24/16
25		25/16
26	A[c]	26/16
27		27/16
28	B[a]	28/16
29		29/16
30	B	30/16
31		31/16
32	C₅	32/16
33		33/32
34		34/32
35		35/32
36	D	36/32
37		37/32
38		38/32
39		39/32
40	E	40/32
41		41/32
42		42/32
43		43/32
44	F[b]	44/32
45		45/32
46		46/32
47		47/32
48	G	48/32
49		49/32
50		50/32
51		51/32
52	A[c]	52/32
53		53/32
54		54/32
55		55/32
56	B[a]	56/32
57		57/32
58		58/32
59		59/32
60	B	60/32
61		61/32
62		62/32
63		63/32
64	C₆	64/32

[a] This ratio represents the Just tone B (15/8) lowered by a frequency ratio of 1/8 to give 14/8.

[b] This ratio represents the Just tone F (4/3) raised by a frequency ratio of 1/24 to give 11/8.

[c] This ratio represents the Just tone A (5/3) lowered by a frequency ratio of 1/24 to give 13/8.

with each new octave of intervals. It should be evident from the above discussion that each musical interval in the harmonic series is unique only if the frequency ratio is a prime number. Prime numbers represent new tones, while the other numbers are octave duplications of lower harmonics. For example the frequency ratios 2:1, 4:2 and 64:32 all represent the same musical interval, overtones of C. However, prime numbers like 2:1 and 3:2 give unique musical intervals of the octave and fifth, respectively.

One way to conceptualize these harmonics is to think of the frequencies of the overtones as different segments on a piece of string. For example, let the tone C, with a frequency ratio of 514 vibrations/second, have the string length equal to one. To construct the tone D in the harmonics series, which has a frequency ratio of 9/8, we would need to shorten the string by 1/9th of the original string length. The remaining 8/9th of the string would represent the tone D. The greater the number of harmonics the smaller the differences in string length that will be needed in order to distinguish the tones (Helmholtz, 1954).

In theory, the harmonic overtone series extends to infinity, but in practice, a musical tone is supported by a limited number of overtones. In Figure 10 we have represented the first sixty-four harmonic overtones. A whole new array of notes can be derived from the harmonic series responsible for the richness that we hear in musical tones. If these overtones were not present, then a tone would be characterless. A more detailed discussion of musical intervals is given in Chapter 14 where we relate these fundamental concepts of music theory to the numerical pattern responsible for the genesis of the material universe.

The Harmonic Structure of a Musical Tone Matches the Constitution of the Universe

At this stage we have presented a basic understanding of the musical tone and the harmonic overtone series at the foundation of music theory. Based on these principles, higher musical structures are built which give the musician the diverse aesthetics of melodies, rhythms, tunes and other musical aspects. Also, using these principles of music theory, we can draw a complete correspondence with the Constitution of the Universe (see Chapter 1). The first unit in the Constitution of the Universe is repre-

sented in Rig Veda by the syllable, "A-K." According to the Vedic
Literature, "A" represents the state of all possibilities, which collapses to
the point value of "K." Contained within "A-K" is the seed form of all
knowledge in creation. Similarly, from the bed of silence ("A") arises a
musical tone ("K"). Inherent in the structure of a musical tone are all tonal
relationships, musical intervals and values of consonance and dissonance.
Often a single tone or key note (called a tonic) establishes the central
point around which a musical piece revolves. It is analogous to a traveler
who thinks of each point on his or her trip in terms of the distance from
"home." Similarly, in music the notes in a tune or melody are considered
in terms of their distance from the key note and we often anticipate the
piece ending on that note.

The next level in the Constitution of the Universe is the unfoldment of the
first syllable "A-K" into the eight syllables or prakritis of the first pada
(refer to Figure 1). Likewise, from a single musical tone, the first and most
basic interval in the harmonic overtone series to be constructed is the
octave. For most scales, the octave interval contains a series of eight notes
which, we propose, correspond to the eight prakritis of the Constitution
of the Universe. Table 8 lists the eight musical tones and their matching
prakritis. Apart from the mere quantitative correspondence between the

---

Table 8. Correspondence Between the Eight Prakritis of the Constitution of the Universe and the Octave of Eight Tones with their Associated Emotional Qualities

Rig Veda:

AK	ni	mi	le	pu	ro	hi	tam

Prakriti:

Ahamkara	Buddhi	Manas	Akasha	Vayu	Agni	Jala	Prithivi

Translation:

Ego	Intellect	Mind	Space	Air	Fire	Water	Earth

Musical Tone:

C (Do)	D (Re)	E (Mi)	F (Fa)	G (So)	A (La)	B (Ti)	C (Do)

Emotional Quality:

Strong, Firm	Rousing, Hopeful	Steady, Calm	Desolate, Awe-inspiring	Bright, Grand	Sad, Weeping	Sensitive, Piercing	Strong, Firm

eight notes and the eight prakritis, there also appears to be a deeper significance. Some musicians have purported that each individual note has an associated emotional quality. The following parallels are suggested between the first eight syllables or prakritis of Rig Veda and a list of notes and their emotional qualities taken from Curwen's *Standard Course of Lessons and Exercises in the Tonic Sol-fa Method* (Jeans, 1968, p. 184):

- C (Do) is associated with the feeling of being "strong" or "firm" relating to the first prakriti, Ahamkara (translated as ego). Ego is the innermost subjective quality and is very close to the level of pure consciousness. At this level, it exemplifies stability, strength and firmness which give integrity to our life.

- D (Re) is associated with the feeling of "rousing" or "hopeful" which matches the second prakriti, Buddhi (translated as intellect). The qualities of the intellect are those of discrimination, reasoning and thinking. The feeling of rousing often involves becoming excited or awake which may be associated with the active nature of the intellect. In addition, the intellect has the ability to reason or justify which often gives us hope and encouragement.

- E (Mi) is associated with the feeling of "steady" or "calm." These emotional feelings correspond to the qualities of the third prakriti, Manas (translated as mind). The mind may be described as that entity which perceives and feels. For the mind to exhibit awareness, memory and intelligence, steadiness and calmness naturally exist to some extent. Metaphorically, the mind is also often associated with a lake, embodying the feelings of stability, steadiness and calmness from which thoughts "bubble up." (It is of interest to note that in the I Ching (refer to Chapter 8, Section IV) this third prakriti corresponds to the trigram "Tui" which is translated as "lake.")

- F (Fa) is associated with the feeling of "desolate" or "awe-inspiring." These qualities match the characteristics of the fourth prakriti, Akasha (translated as space). The feeling of "desolate" is particularly apt because space is often viewed as a huge arena of emptiness, blackness, and, except for the occasional star cluster, devoid of life. Yet at the same time humanity has, throughout the centuries, gazed into the heavens with awe, inspiration and wonder.

- G (So) is associated with the feeling of "bright" or "grand"
  corresponding to the fifth prakriti, Vayu (translated as air).
  Essentially air is clear, lucid and a medium through which light
  travels. Hence the feeling of "bright" is apt. Air also occupies a
  huge space which could evoke the feeling of grandiose.

- A (La) is associated with the feeling of "sad" or "weeping"
  matching the sixth prakriti, Agni (translated as fire). Fire may
  produce destruction, death and loss of cherished objects. Such
  properties can often provoke weeping, feelings of sadness or
  depression.

- B (Ti) is associated with the feeling of "sensitive" or "piercing"
  matching the seventh prakriti, Jala (translated as water). These
  feelings may be related to the physical qualities of water. Water
  is usually sensitive to any perturbations which create turbulence
  and ripples. In addition almost any object is able to penetrate
  or pierce this medium.

- C (Do) is associated with the feeling of being "strong" or "firm."
  These two feelings correspond to the qualities of the eighth
  prakriti, Prithivi (translated as earth). Earth provides the
  foundation or support on which we have built civilization.
  It provides strength, firmness and stability.

From these eight syllables of the first pada of Rig Veda, the next level of
expansion in the Constitution of the Universe is the interpretation of each
of these eight prakritis with respect to rishi (knower), devata (process of
knowing) and chhandas (known) producing a total of twenty-four units.
Likewise, in music theory each tone is distinguished from other tones in
three ways: quality, pitch and force (Helmholtz, 1954). The first characteristic describing a musical tone, quality, is the distinctive feature we hear
between instruments which play sounds of the same volume and pitch.
This aspect of a musical tone corresponds to the nature of rishi which
equates with the inherent or abstract nature of a tone. The second way to
describe a musical tone, pitch, is simply the frequency or cycles/second of
that sound. Frequency implies movement, dynamics and action which are
exactly the qualities embodied by the devata interpretation. The final
distinguishing quality of a musical tone, force, is a measure of the
loudness of a particular sound. This quality is equated with the chhandas
interpretation which expresses the physical or material nature of the tone.
The force behind a tone produces the actual physical sensation of a

musical tone. These three disparate peculiarities for each tone of the octave of eight notes comprise a total of twenty-four units corresponding to the twenty-four units of the Constitution of the Universe.

In the next stage of elaboration of the Constitution of the Universe the eight syllables of the first pada of Rig Veda expand into sixty-four units. In music theory we propose that this corresponds to the harmonic overtones series which build upon the octave interval. Since the construction of the harmonic overtones follows a binary sequence, there is a stage in which there will be exactly sixty-four overtones. These sixty-four overtones cover most of the practical range of notes used in the harmonic series. Although in theory other overtones are possible beyond this level, there comes a point where the intervals of the higher harmonics become so close that they are eventually indistinguishable by the human ear. Thus, we feel that the first sixty-four harmonics for a musical tone match the sixty-four syllables of the Constitution of the Universe.

The final correspondence with the Constitution of the Universe stems from the sixty-four basic syllables of Rig Veda interpreted with respect to rishi, devata and chhandas, thus comprising a total of 192 units. Likewise each of the first sixty-four harmonic overtones of a musical tone can be viewed in relation to the three peculiarities of each tone (thus a total of 192 units).

The Harmonic Overtone Series is Related to the Periodic Table and the DNA

There are several fascinating parallels between the harmonic structure of music and the fundamental aspects of other fields. First, in the discipline of chemistry, authors have often noted an intriguing similarity in the structure of the periodic table of elements and the octaves of music. For example, in 1864 the English chemist John Newlands proposed a structure for the arrangement of elements patterned after the notion that particular chemical characteristics are found in every eighth element, similar to the musical tones repeating every octave (Zumdahl, 1989). Later, Dmitri Mendeleev improved on the work of Newlands and others to create the periodic table as we know it today. As discussed earlier in Chapter 3, Section I, the structure of the electron configurations in the periodic table of elements parallels exactly the structure of the Constitution of the Universe. It is interesting that the visual structure of the periodic table

(Figure 4; e.g., increasing number of electrons at each new orbital) looks very similar to the pattern of harmonic overtones in which each new level has an increased number of tones (Figure 10).

In the field of biology, a Japanese geneticist, Dr. Susumu Ohno, in the mid-1980s published several papers on "gene music." What first led Dr. Ohno to postulate a link between the genetic code and music was the idea of redundancy or repeating sequences (Ohno and Ohno, 1986). Redundancy in music is displayed in the recurrence of a basic tune or melody and in DNA as the existence of multiple copies of particular genes. Developing this idea, Dr. Ohno proceeded to establish a rule for transforming DNA coding base sequences into musical score, and vice versa. Each of the four nucleotide bases (adenine, guanine, thymine and cytosine) were assigned two consecutive notes on the octave scale in ascending order (Ohno and Jabara, 1985). Although this work has produced promising developments (e.g., cancer genes sounding "hauntingly melancholy" like funeral music), additional research is needed to remove the subjectivity which is often needed to transform a gene sequence into a musical score. Perhaps further development in this area could be catalyzed by studying the similarities between the Constitution of the Universe for the genetic code of DNA (Table 7) and the harmonic overtones of music (Figure 10). For example, we could start by assigning each note in the octave to each of the eight codon-codon base pairs of the DNA double helix. In this way each codon (set of three base pairs) would correspond to one particular note. Whether this "new" rule for transforming DNA base pairs into musical scores would be valuable, remains an area of further research. These two examples of the relationship between music and the disciplines of chemistry and biology signify the potential for numerous developments which could emerge through the study of universal or general principles linking one discipline to another.

Chapter 6:

Art and the Constitution of the Universe

Section I

The Science of Color

We live in a universe where everything consists of, and is distinguished by, color. As a result, it is not surprising that the use of color has been a major avenue of expression for artists. Although artists have intuitively understood how to use colors to create harmony and beauty, they have generally avoided classifying colors into systematic and orderly systems. Consequently, even the greatest artists have not completely understood the color phenomena. Thus, most of our current knowledge of color theory is attributed to the work of scientists. In this section we will describe the "grammar rules" of color based on the work of the leading color theorists. Further, we will show that the structure and order of the most well-accepted color systems match the same numerical pattern of the Constitution of the Universe.

Origins of Color Theory

Historically, Sir Isaac Newton was the first scientist accredited with examining color as a unique property of matter. In 1667 he demonstrated that a prism separates white light into a rainbow of colors ranging from red to violet. Newton, however, was concerned with just the pure colors of the spectrum and did not deal with black-containing or gray-containing colors (Birren, 1969b). During the centuries following Newton's work, numerous developments were made in constructing comprehensive color systems. The two most important and best-known color systems to emerge from these endeavors are based on the work of Albert Munsell and Wilhelm Ostwald. Interestingly, what we have discovered is that the color system developed by Wilhelm Ostwald reflects the numerical structure of the Constitution of the Universe (presented in this section) and the color system of Albert Munsell reflects the numerical pattern of the genesis of the material universe (refer to Chapter 15). It is fascinating to note that both Ostwald and Munsell knew each other, and there are many basic similarities between their color systems.

Dr. Wilhelm Ostwald, a Nobel Prize winner in chemistry, was one of the few scientists to devote a large portion of his time to understanding the problems of color theory. In academic circles Ostwald's color system has gained acceptance and prestige in England, Europe and the United States. Also, many commercial industries (e.g., textiles, wallpapers, carpeting, etc.) have relied heavily on the principles developed by Ostwald.

Definition of Color

To begin our discussion of Ostwald's color system, we need to define color. Like many other common words in the English language, color has a variety of meanings. The physicist may describe color as an electromagnetic wave within the color spectrum. Using this definition, we find an infinite number of possible colors. To the chemist, color is a unique property of matter and is seen as a pigment on the artist's palette, a dye or a coloring agent. But color appears to be something much deeper than what these definitions suggest. When we perceive color, it is a highly personal, subjective experience – what one person calls a particular color may not be the same as what another person calls that same color. In this light, it is reasonable to think of color as a *sensation* (Birren, 1969a; Birren, 1969b). Every color enlivens a different emotion and aspect of our consciousness. It is as if color is structured in consciousness. Defining color as a sensation automatically implies that black and white are colors because, when we see either one of these colors, we get some type of real sensation. To think of white as all colors, and black as the absence of colors, has no real meaning when color is viewed as a sensation.

There are two broad classes of colors: achromatic and chromatic. Achromatic colors consist of white, black and all variations of gray. Chromatic colors consist of all other colors (e.g., blue, green, red, yellow, etc.). All colors, chromatic and achromatic, can be described in three distinct ways (Nickerson, 1946; Birren, 1969a). The first, the hue of a color, describes the quality which distinguishes one color from another (i.e., the name of the color). For example, a color may be forest green, blue or orange. The second way to describe a color, the value or lightness of a color, denotes whether a color is light or dark. For example, a light red color is referred to as pink and a dark red may be called maroon. The last way to describe a color, by its chroma or intensity, describes the amount of gray in the color which determines whether the color is dull or vivid.

To understand the variety of all these colors in an orderly and systematic array, Ostwald constructed what is known as a color solid (Jacobson, 1937; Birren, 1969b). Figure 11 presents a view of the fully constructed Ostwald color solid.

We will now describe each component of the Ostwald color solid, enabling us to give order and understanding to the world of colors.

The "Gray Scale" of Achromatic Colors

The central axis of the Ostwald color solid, called the "gray scale," is composed of the achromatic colors. A gradation of gray colors is arranged between the two extreme endpoints of black and white. Theoretically the gradations from black to white are infinite because between any two grays it is possible to insert a third gray which is lighter than one and darker than the other. Eventually, however, the differences in the divisions become imperceptible to the human eye. Ostwald found that there were eight gradations from white to black that were near enough together so that intermediates would be superfluous and far enough apart to be readily distinguished (Jacobson, 1937).

Each of these eight gradations from white to black is represented by a single letter denoting the relative percentages of their white and black content. For each of these achromatic colors the total amount of white and black will be equal to 100%. The following summarizes the eight gradations:

Letter:	a	c	e	g	i	l	n	p
% White:	89	56	35	22	14	8.9	5.6	3.5
% Black:	11	44	65	78	86	91.1	94.4	96.5

It is important to note that it is not possible to obtain a "perfect" white and a "perfect" black color. By measuring brightness (the amount of light

Figure 11. The Ostwald Color Solid

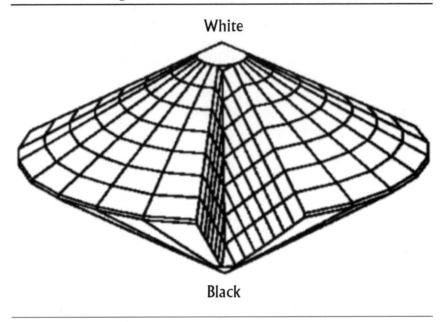

Adapted from Birren, F. (1969b). *A Basic Treatise on the Color System of Wilhelm Ostwald: The Color Primer.* New York: Van Nostrand Reinhold Company, p. 18.

reflected by an object) Ostwald found that the blackest object (which theoretically absorbs all light) contains 3.5% white and that the whitest object (which theoretically reflects all light) contains 11% black. For these reasons, the first gradation in the gray scale, represented by the letter "a," contains 89% white. Similarly, the final step in this scale, symbolized by the letter "p," contains 96.5% black.

The series of gray colors between white ("a") and black ("p") are visually equidistant to one another, and the eye perceives the difference in the gradations as being equal. In actuality however, the percentage of white content in the gray scale decreases geometrically because the addition of large amounts of black causes only faint visual changes among the gray colors whereas the addition of the slightest quantities of white causes a substantial lightening. This is evident by examining the percentage differences at both ends of the gray scale – addition of 33% black at the white end produces the same difference in gradation as the addition of 2% white to the black end (Birren, 1969b). Therefore, it appears reasonable to arrange the gradation of the gray scale according to our visual judgment as to what appear to be equal intervals. If we were to arrange the gradations according to the content of white or black we would end up with too many divisions at the white end and too few at the black end (Birren, 1969b).

The "Hue Circle" Contains Eight Fundamental Colors

Just as the achromatic colors made up the gray scale, Ostwald proceeded to create what is known as the "Hue Circle," which forms the outer perimeter of the Ostwald color solid (refer to Figure 12). This Hue Circle consists of pure chromatic colors. In theory this circle is continuous, meaning that there are an infinite number of chromatic colors, because between any two different hues it is always possible to place a third one. As with the gray scale, however, the divisions between chromatic colors continue until the differences become imperceptible to the human eye. Ostwald found that a total of twenty-four chromatic colors were near enough together so that intermediates would be superfluous and yet far enough apart to be readily distinguishable. These twenty-four colors were considered to be sufficient for practical purposes and were adopted as the color standard.

According to Ostwald, the twenty-four chromatic colors of the Hue Circle are formed by eight principle or fundamental colors (refer to Figure 12). These eight basic colors are as follows: yellow, orange, red, purple, blue, turquoise, sea green, and leaf green. For each of these eight fundamental colors Ostwald added intermediate colors until he formed the complete circle of twenty-four colors. There are three hues associated with each of the eight fundamental colors (Birren, 1969b).

In contradistinction to the gray scale of achromatic colors, which is a linear sequence, the series of twenty-four chromatic colors is arranged in a circle. The starting point in the circle is arbitrarily chosen as yellow and proceeds in the direction of red. The twenty-four chromatic colors are numbered one through twenty-four (refer to Figure 12). The twenty-four colors form themselves into a circle because, if we start at any given color, the sequence of colors which follow become increasingly different and then start to become increasingly similar until we reach the same color that we began with. Thus, for every color in the circle, there is another

Figure 12. Eight Fundamental Colors Compose the
Twenty-Four Divisions of Ostwald's Hue Circle

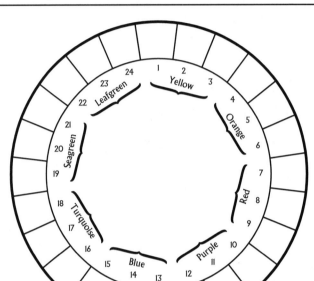

Adapted from Birren, F. (1969b). *A Basic Treatise on the Color System of Wilhelm Ostwald:
The Color Primer.* New York: Van Nostrand Reinhold Company, p. 38.

color which is the most different from or least similar to it. Such a pair of
colors is called "complementary," and the colors are located opposite one
another in the Hue Circle, for example, yellow/blue, purple/leaf green,
orange/turquoise and red/sea green. When complementary colors are
mixed together they form neutral gray.

The Monochromatic Triangle Displays
All the Variations of a Chromatic Color

For each of the twenty-four chromatic colors Ostwald proceeded to
develop a monochromatic triangle which has become recognized as the
single, most remarkable feature of Ostwald's color system (Birren, 1969b).
Each monochromatic triangle consists of all derivatives of one chromatic

color (hence the word "monochromatic") formed by mixing that hue with the achromatic colors of the gray scale. All variations between the full chromatic color located on the circumference of the Ostwald color solid and the neutral gray axis find their place in the construction of the monochromatic triangles.

In Figure 13 we have reproduced the typical monochromatic triangle which can be constructed for each of Ostwald's twenty-four chromatic colors or hues. Each of these triangles consists of twenty-eight different colored rhombuses plus eight neutral grays. The three corner rhombuses are formed by the pure chromatic color, white and black. On the right side of the monochromatic triangle are the gradations from white to black constituting the gray scale (see Figure 13). As described earlier, each of the eight gradations in the gray scale is represented by a single letter which denotes its content of black and white. The other twenty-eight colored rhombuses are denoted by two letters: the first letter tells the white content of the color and the second letter specifies black content. To calculate the amount of pure color in each rhombus, we need to use the following formula (Jacobson, 1937):

100% - (% of white + % of black) = % pure hue content

Thus, a pure chromatic color is represented by "pa," meaning that it contains no black and no white. (Note: since there is no such thing as a "perfect" white or black, a pure color does actually contain 11% black and 3.5% white.) The percentages of pure color, black and white can be calculated accordingly for each rhombus in Figure 13. For example, the rhombus "ne" is a color which is made from 5.6% white ("n"), 65% black ("e") and 29.4% of a pure chromatic color.

On the upper side of the monochromatic triangle (from "pa" to "a") is what is called the Light Clear series. All colors in this row contain no black (or the minimal amount possible, 11%). The first color in the series, the pure chromatic color, gradually becomes lighter and lighter with each color in the series until the last color is reached which is white. A similar row is located on the lower side of the monochromatic triangle (from "pa" to "p") and is called the Dark Clear series. All colors in this row contain the minimal amount of white, 3.5%. This series begins with the pure chromatic color like the Light Clear series, but the colors become gradually darker and darker until the last color is reached which is black. All colors between these two series are specific combinations of the pure

Figure 13. A Monochromatic Triangle Contains Twenty-Eight Colored Rhombuses plus Eight Neutral Grays

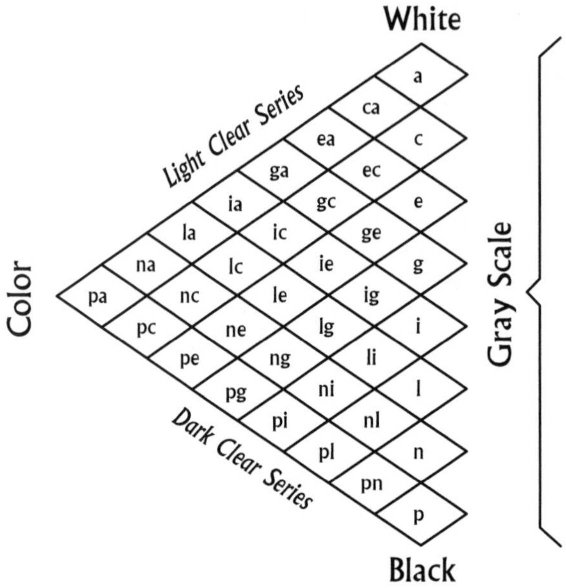

White

Light Clear Series

Dark Clear Series

Color

Gray Scale

Black

Letter	% White	% Black
a	89	11
c	56	44
e	35	65
g	22	78
i	14	86
l	8.9	91.1
n	5.6	94.4
p	3.5	96.5

- **Colors are Represented by Two Letters:**
  First letter = % white content
  Second letter = % black content
  100% − (white content + black content) = % hue content

- **For Example:**
  "ne" is a color which has 5.6% white, 65% black and 29.4% pure hue

Adapted from Jacobson, E.G. (1937). *The Science of Color: A Summary of the Theories of Dr. Wilhelm Ostwald.* St. Louis, Missouri: Barnes-Crosby Company.

chromatic color with some amount of gray. For a clear visual picture of the monochromatic triangles the reader is referred to the work of Jacobson (Jacobson, 1937) where all diagrams are in color.

For each of the twenty-four chromatic colors of the Hue Circle, a monochromatic triangle is constructed. These triangles are then arranged like spokes around a wheel to form the complete Ostwald color solid. Each monochromatic triangle uses the same notational system. Thus, to distinguish between different triangles, the assigned number of the chromatic color is placed in front of all letters for that particular triangle. For example, the notation "1-ne" represents a color which has 5.6% white, 65% black and 29.4% yellow (the number for yellow is 1).

A Summary of the Ostwald Color Solid

We have now described all the components of the Ostwald color solid which display the entire world of color that can be used and combined by the artist to create balance and harmony. Referring back to Figure 11, we can see where all the components fit together. When all the components are combined, the final three-dimensional solid is the shape of a double cone. The central axis of this solid is the gray scale and the chromatic colors of the Hue Circle are located on the circumference. The upper surface of the double cone contains the lighter colors, while the lower surface contains the darker colors. Toward the center of the solid, the colors become increasingly grayer until the gray axis of achromatic colors is reached. The peaks of all the monochromatic triangles converge in the top of the solid at white and in the bottom of the solid at black. All colors distinguishable by the human eye are represented in this color solid. Ostwald considered it to be the final solution to the problems of color theory for the entire world of colors was displayed in it (Birren, 1969b; Jacobson, 1937).

If we take a "slice" through the Ostwald color solid, we obtain what is shown in Figure 14, essentially two monochromatic triangles joined together. The two hues from which the monochromatic triangles are derived in any cross-section are complementary to each other. In fact, all the pairs of twenty-eight colors in both monochromatic triangles of a cross-section are complementary to each other. As stated earlier, when two complementary colors are mixed, we obtain a neutral gray, which is located in the center axis of the diagram. Thus, each cross-section in the Ostwald color solid contains all the color variations comprising a set of complementary colors. The total number of rhombuses in any such cross-section is equal to sixty-four (twenty-eight colors from one hue + twenty-eight colors from the second or complementary hue + eight colors of the gray series).

Figure 14. A Cross-Section of the Ostwald Color Solid:
Total Number of Rhombuses Equals Sixty-Four

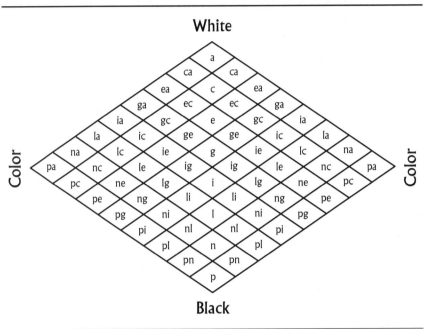

Adapted from Jacobson, E.G. (1937). *The Science of Color: A Summary of the Theories of Dr. Wilhelm Ostwald.* St. Louis, Missouri: Barnes-Crosby Company.

Ostwald's Color Theory is Identical to the Constitution of the Universe

At this stage, we are now able to use the basic principles of the Ostwald color system described in this section to draw a complete correspondence with the Constitution of the Universe. Our first connection relates to the first level of unfoldment in the Constitution of the Universe template, from the first syllable "A-K" into the eight syllables or prakritis of the first pada (refer to Figure 1). We propose that the eight principle or fundamental colors on which Ostwald based his Hue Circle of twenty-four colors (refer to Figure 12) match the eight prakritis of the Constitution of the Universe. It is also interesting to note that Leonardo da Vinci also claimed that there were only eight colors which Nature produces: yellow, red, purple, blue, tawny or umber, green, black and white (Da Vinci, 1802, p. 117). Another fascinating point is that the wavelengths of the visible light spectrum span from 400 nm to 800 nm – from red to violet. This interval is approximately a doubling in the wavelength (also known as an octave). Clearly, there seems to be a fundamental eight element of color underlying the discipline of art.

From the eight basic syllables of the first pada of Rig Veda, the next level of expansion in the Constitution of the Universe is the interpretation of each of the eight syllables with respect to rishi (knower), devata (process of knowing) and chhandas (known), producing a total of twenty-four units. Similarly, any color can be described in three basic ways which we can match satisfactorily with rishi, devata and chhandas. The first characteristic, hue (quality), distinguishes one color from another color. It is the name of the color (e.g., ocean blue, brown, etc.). This aspect of color corresponds with the nature of rishi which is equated with the inherent or abstract nature of an entity. The second quality of a color, the value or lightness, describes whether a color is light or dark (e.g., ocean blue vs. navy blue). This aspect of color, we propose, corresponds to the nature of devata. The last characteristic of a color, chroma or intensity, denotes the strength of a color or the amount of gray it contains. We propose that this last characteristic of a color matches the nature of chhandas.

In the next stage of elaboration of the Constitution of the Universe the eight syllables of the first pada of Rig Veda expand into sixty-four units. A similar structure can also be located in Ostwald's color solid. As described earlier, if we take any "slice" or a cross-section of the Ostwald color solid, we will obtain an array of sixty-four rhombuses. These sixty-four rhombuses describe all the color variations for any pair of complementary colors. Down the center of the sixty-four array is the gray scale of eight achromatic colors and on either side are the monochromatic triangles for the pair of complementary colors.

The final correspondence with the Constitution of the Universe is the interpretation of each of the sixty-four aspects with respect to rishi, devata and chhandas. This creates a total of 192 units. Likewise each of the sixty-four colors in a cross-sectional view of the Ostwald color solid can be viewed with respect to the three characteristics of a color, namely, hue, value and chroma (a total of 192 units).

The Constitution of the Universe is also Found in Crayola Crayons!

In concluding this section, we would like to give a delightful example of finding the numerical structure of the Constitution of the Universe in a successful system of applied color, that from the Binney and Smith company, famous for their Crayola Crayons. For almost a century this company has distributed their famous box of eight crayons. Interestingly the same eight colors which were in the original box in 1903 are still used today. Then, in 1958 Crayola manufactured a "classic box of 64" crayons which was the ultimate in crayon assortments for children. Recently several larger boxes have appeared on the market which hold 96 (12 × 8) and 112 (14 × 8) crayons.

From the consideration of numbers alone, eight has obviously been a number of prime significance in Crayola history. Further, it seems more than just a coincidence that the number of crayons in their classic boxes of eight and sixty-four exactly match the key numbers of the Constitution of the Universe. In Table 9 we have summarized the crayon colors in

---

Table 9. Summary of the Correspondence Between the Constitution of the Universe and the Crayon Colors Chosen by Crayola Crayons

---

Crayon Colors found in the Classic Box of Eight:

Black	Brown	Orange	Yellow	Green	Blue	Violet	Red

Crayon Colors found in the Classic Box of Sixty-Four:

Black	Brown	Orange	Yellow	Green	Blue	Violet	Red
Gray	Sepia	Red Orange	Yellow Green	Blue Green	Midnight Blue	Orchid	Violet Red
Cadet Blue	Raw Sienna	Bittersweet	Green Yellow	Aquamarine	Navy Blue	Mulberry	W. Strawberry
Periwinkle	Burnt Sienna	Burnt Orange	Spring Green	Jungle Green	Cerulean	Red Violet	Magenta
Silver	Mahogany	Tan	Golden Rod	Pine Green	Cornflower	Fuchsia	Salmon
Gold	Indian Red	Melon	Dandelion	Forest Green	Teal Blue	Plum	Carnation Pink
Copper	Brick Red	Peach	Yellow Orange	Sea Green	Turquoise Blue	Royal Purple	Lavender
White	Maroon	Apricot	Vivid Tangerine	Olive Green	Sky Blue	Blue Violet	Thistle

Each Crayon Color is Described in Three Ways:

- The Hue aspect of the Color   *(Rishi Interpretation)*
- The Value aspect of the Color   *(Devata Interpretation)*
- The Intensity aspect of the Color   *(Chhandas Interpretation)*

Crayola's box of eight and sixty-four. The crayon colors in the box of eight have only two colors different from the eight principle colors of Ostwald and only one different from the eight chosen by Leonardo da Vinci.

It is tempting to suggest that the great success of Crayola Crayons may be due in part to the number of crayons which they selected to use in their classic boxes. During this century, Crayola Crayons have become a household word and have gained a long lasting history of acceptance by millions of children.

SECTION II

VEDIC ARCHITECTURE

In ancient civilizations the design of temples and other such enclosures reflected the image of the universe and represented a synthesis of cosmic order. Buildings were constructed using well-defined principles aimed at creating balance and grace. Often these sites displayed a mosaic of sacred symbols indicating that they were a link between God and man, a dwelling place for the Supreme Spirit on Earth. Each enclosure was a microcosmic blueprint of the macrocosm. Unfortunately, over the lapse of time our modern era has lost most of this ancient wisdom of architecture. As a result, much of what we call architecture today is devoid of grace and often dissatisfying. In this section we will present a detailed discussion of ancient Vedic architecture. This particular example of ancient architecture was chosen because holistic knowledge is available on the topic. Further, the fundamental principles of ancient Vedic architecture are a clear reflection of the numerical structure of the Constitution of the Universe. At the end of this section we will briefly summarize examples from other major civilizations which also match the structure of this universal template.

The Vastupurusamandala is the Foundation of Vedic Architecture

The traditional science of Vedic architecture is "vastu-sastra," translated as the rules governing architecture. It deals with all aspects of architecture from measurements, to site selection, to orientation of the building, to materials used in construction and to the planning of villages, towns and temples. Although this science is recorded in various scriptures, the knowledge came from the undefined past by way of oral traditions. The earliest datable source on vastu-sastra is Brhat Samhita which is considered to be one of the authoritative texts on the subject. In the Vedic civilization there were four classes of craftsmen: a sthapati (architect-priest), a sutragrahin (surveyor), a taksaka (sculptor) and a vardhakin (builder-plasterer-painter). Of these four, the sthapati or Vedic architect-priest was considered the foremost. All other craftsmen carried out the instructions of the sthapati (Kramrisch, 1976).

According to the Brhat Samhita, all temples, buildings, villages, markets, cites, capitals and other enclosures were constructed on a prototype called the "vastupurusamandala." Knowledge of the vastupurusamandala was the first discipline a Vedic architect had to master. There is no text on Indian architecture which does not deal with the principles of the vastupurusamandala or which takes its knowledge for granted. The word vastupurusamandala consists of three parts: vastu, purusa and mandala. "Vastu ... is the extent of Existence in its ordered state and is beheld in the likeness of the Purusa." (Kramrisch, 1976, p. 21). Purusa is the totality of manifestation, the seed image of Brahma and is known as the primordial or cosmic man. In his image or likeness the site plan of the building was constructed. The word mandala refers to the physical plan of the building. Although a mandala can denote any closed polygon, the shape of a square was considered to be the most essential form, from which other forms could be derived (Kramrisch, 1976).

The vastupurusamandala is a sacred geometrical diagram representing the structure of the universe and was used to construct all architectural forms. It was a diagram or a forecast of what the final building would look like. All aspects and contents of an enclosure were predetermined based on the vastupurusamandala. The vastupurusamandala, however, was not a ground plan; rather, it consisted of schematic principles which regulated all architectural forms. Building in accordance with the principles of the vastupurusamandala allowed the gods to dwell at these sites in peace. Before a building was constructed, the ground was leveled and consecrated, and the vastupurusamandala was drawn on the site. In this way the building extracted from the Supreme Force residing at the foundation of the architectural form. The presence of the Supreme Principle transformed the mandala plan into the physical structure of the building.

The Most Sacred Vastupurusamandala is a Diagram of Sixty-Four Squares

In brief, the vastupurusamandala was a simple grid or square mandala which might be of any size and could be converted into other primal shapes. Although the shape of a mandala might also be circular, the square was considered to be a mark of order and of greater significance. Moreover, of all architectural designs the square was considered to be the leading symbol and was literally the foundation of Vedic architecture

(Kramrisch, 1976). Contained in the shape of a square were the cycles of measurable time and knowledge of cosmic order. The square was typically used in the vastupurusamandala because of its relation to the proportions of the human figure. On average, the span of an individual's outstretched arms is equal to his/her height. With respect to a square this means that the height of a figure and length of its outstretched arms, each equal the length of one edge of the square. These same ideas were also known to Leonardo da Vinci (refer to Chapter 7, Section I). Thus, by using a square mandala, the Vedic architects represented the proportions of man and the cosmos in their designs.

Each vastupurusamandala was designed by subdividing the total square mandala into smaller squares, all identical to one another. Although there were a number of various possible designs, the Brhat Samhita mentions two prominent mandala plans consisting of sixty-four and eighty-one squares. Of these two plans, the structure of sixty-four was considered to be the more sacred because it reflected the order which brought the universe into existence (Kramrisch, 1976). This design was used especially in the construction of shrines, temples and other holy places of worship. Further, it was known that the structure of sixty-four was included in the mandala of eighty-one squares.

Organization of the Vastupurusamandala Containing Sixty-Four Squares

Within each matrix of sixty-four compartments various deities were assigned to particular squares. One of the most prominent features in all buildings was the importance ascribed to the center four squares. These central squares, the "Brahmasthana" of the building was the seat of the primal deity Brahma, known also as the embryo of splendor or the primordial germ of cosmic light. The four squares forming the Brahmastana appear to correspond to the four Vedas (Rig, Sama, Yajur and Atharva). Around the nucleus of the Brahmastana were located two sets of auxiliary deities. Of the twelve deities assigned to the first set, eight were given special significance. These eight deities were called the eight vastupurusas and corresponded to the eight cardinal directions (NE, E, SE, S, SW, W, NW, N). Using the eight directions in measurement, the architect was able to ascertain the correct position of the building in the cosmos. The second set consisting of thirty-two deities were associated with the celestial bodies (twenty-eight were lunar and four were related to the solstical and equinoctial points). These thirty-two deities were located on the outer border of the vastupurusamandala. (Note: there are only twenty-eight squares in the outer border for a sixty-four square mandala. The extra four compartments were created by dividing each of the corner squares by diagonal lines.) The thirty-two deities provided astronomical connotations regarding measurement of the days, months, years, etc. For this reason the Vedic science of architecture was closely related to the discipline of astronomy (Kramrisch, 1976).

In total there were forty-five deities represented in each vastupurusamandala: one Brahma, twelve inner and thirty-two outer. By arranging these deities in various patterns, the Vedic architect had a range of possibilities in the design of architectural forms. However, one thing in common was the location of the Brahmasthana in the center, and the influence of all other deities was arranged with respect to the Brahmasthana.

Once the vastupurusamandala was designed, it became possible to construct the physical form of the building (for more details refer to Kramrisch, 1976). This final phase of construction was a time of great artistic expression and diversity. Often the exterior walls of the temples were richly embedded with sculptures and carvings. Interestingly, it appears that measurements of vertical and horizontal aspects of the building usually matched the octave and fifth intervals of music. A properly built enclosure ensured growing prosperity, health and harmonious relations. Any visitor entering a building constructed on the vastupurusamandala was said to become recharged with cosmic influences and may even have forgotten, for a moment, the purpose of the visit (Das, 1989).

The Vastupurusamandala and the Constitution of the Universe

Based on this overview of fundamental principles involved in Vedic architecture, we will now proceed to locate in the structure of the vastupurusamandala the matrix of the Constitution of the Universe. Our first correspondence relates to the first unit of the Constitution of the Universe, "A-K." According to Rig Veda, "A" represents a state of all possibilities collapsing to the point value of "K." Likewise in the vastupurusamandala, the central squares ("K") represent the totality of Brahma ("A"). As described earlier, the central squares formed the Brahmasthana, around which all other deities and principles of architectural form revolved.

The next level in the Constitution of the Universe is the unfoldment of the first syllable "A-K" into the eight syllables or eight prakritis of the first pada (refer to Figure 1). Likewise, the first group of deities surrounding the Brahmasthana in the vastupurusamandala consisted of twelve deities, eight of which had special significance. These eight deities were known as the eight vastupurusas and corresponded to the eight cardinal directions. We propose that these eight deities correspond to the eight prakritis of the Constitution of the Universe.

From these eight syllables of the first pada of Rig Veda, the next level of expansion in the Constitution of the Universe is the interpretation of each of these eight syllables with respect to rishi (knower), devata (process of knowing) and chhandas (known). This produces a total of twenty-four units. Similarly, the name vastupurusamandala consists of three parts – vastu, purusa and mandala – corresponding to the interpretations of rishi, devata and chhandas, respectively. Vastu is associated with the order and intelligence of Existence which is expressed in an architectural form. This matches the rishi aspect which embodies the qualities of abstract knowledge and intelligence. Purusa is the totality of manifestation embodied in the Supreme Spirit. Through the Supreme Spirit the intelligence of vastu is converted into the physical structure of the building. This description corresponds to the nature of devata which is to be an active link between rishi and chhandas. The last part, mandala, depicts the physical dimensions of the architectural form enclosing a building. This matches the characteristics of chhandas which are equated with the most expressed value in the physical properties of an entity. These three aspects of the vastupurusamandala, when viewed with respect to the eight vastupurusa deities, comprise a total of twenty-four units corresponding to the twenty-four units of the Constitution of the Universe.

In the next stage of elaboration of the Constitution of the Universe the eight syllables of the first pada of Rig Veda expand into sixty-four units. We propose that these sixty-four units match the sixty-four squares enclosing the most sacred vastupurusamandala. Within these sixty-four squares were arranged the various deities and other architectural features. The final correspondence with the Constitution of the Universe template stems from the sixty-four basic syllables of Rig Veda interpreted with respect to rishi, devata and chhandas, thus comprising a total of 192 units.

Likewise we can view each of the sixty-four squares in the vastupu-rusamandala with respect to the three aspects of vastu, purusa and mandala (thus a total of 192 units).

Other Illustrations of Architectural Design that Display the Constitution of the Universe

In concluding this section, we would like to give several architectural examples from other ancient civilizations which also match the numerical structure of the Constitution of the Universe. First we will begin by citing some famous examples from Vedic India. In the Ramayana (Bala-Kanda, Canto V) the famous city of Ayodhya was laid out on an 8 × 8 grid, like that of a chess board. A similar example is also found in the legendary city of Krishna called Dvaravati. This city was believed to have been laid out with eight streets crossing eight other streets at right angles.

In ancient China the classical texts of K'ao-Kung Chi describe the layout of the royal capital as being constructed on a chessboard grid of sixty-four squares (Pennick, 1989). Interestingly, in this royal capital the administrative center consisted of four squares, similar to the Vedic Brahmasthana. It has also been recorded that in 741 AD Japan divided its nation into sixty-four provinces (Hanayama, 1960, p. 38).

In Athens the Greek architect Andronicus of Cyrrhus built the octagonal structure known as the Tower of Winds. Also in Athens, the Parthenon has eight pillars on the front side. Eight was an important number in European architecture as well. For example, eight-fold pavement designs are common in Westminster Abbey, Canterbury Cathedral, Xanten Cathedral in Germany and several in Italy (Pennick, 1989).

These are just a few examples of the numerous architectural designs which have used the numbers eight and sixty-four in their structures. There are probably numerous other examples of buildings whose structures reflect the Constitution of the Universe. We feel that this is an area of enormous research, not just for studying ancient civilizations, but for improving and transforming modern architecture.

CHAPTER 7:

ANCIENT WISDOM AND THE CONSTITUTION OF THE UNIVERSE

SECTION I

LEONARDO DA VINCI'S CANON OF PROPORTIONS

Few individuals in history have reached the breadth of understanding and creativity as Leonardo da Vinci. As one of the outstanding men of the Italian Renaissance, he was a scientist, artist, hydraulic and aerodynamic engineer, city planner, imaginative inventor, botanist, geologist and architect. With such a tremendous insight into the laws of nature, it is not surprising that his work has survived the centuries and is still highly respected. In this section we will examine one of Leonardo's masterpieces, the Canon of Proportions, and provide convincing evidence that this work of art and science was formulated upon the same numerical structure as the Constitution of the Universe.

Leonardo da Vinci's Life

From his childhood Leonardo grew up with an intense interest and observation of the natural world. He believed Nature to be a rich source of wisdom and insights. For example, he was fascinated with the flights of birds, the subtle play of shadows and light, and the beauty of flora. In the area of anatomy and physiology Leonardo viewed man as the measure of all things, as a microcosmic image of the world. He believed that understanding the structure of the body was necessary to give expression to its spirit (Reynal and Company, 1967).

To Leonardo, diagrams and illustrations were considered to be the best media by which insights and thoughts could be communicated. They provided simple and efficient ways to convey or explain an idea. In the field of anatomy alone Leonardo constructed well over 300 detailed drawings. Prior to his work, contemporary knowledge in anatomy rarely went beyond the subcutaneous level. However, with the numerous dissections on humans and other creatures, Leonardo revolutionized the knowledge of his time (Mantle, 1978).

The Canon of Proportions is a Symbol of Perfect Balance and Harmony

One of the most famous of his anatomical drawings is the Canon of Proportions. Throughout many disciplines this image has provoked widespread and long-lasting interest. In addition, this image has become a favorite among Sacred Geometrists and others as a gold mine of harmonious proportions and universal symbolism. Our analysis of this drawing with respect to the Constitution of the Universe was inspired by the work of Drunvalo Melchizedek (Melchizedek, 1992; Frissell, 1994) and others. We will now describe this diagram in detail and clearly demonstrate that inherent within it is the numerical structure of the Constitution of the Universe.

In Figure 15 we have reproduced Leonardo da Vinci's Canon of Proportions. This drawing is not edited and represents the original work of Leonardo. One of the prominent features of this image is the human body positioned inside a square and a circle. In Sacred Geometry both the square and the circle are considered perfect shapes which embody universal significance. The square, composed of straight lines portrays firmness and stability, representing Earth. In contrast to this, the circle consists of curved lines, exemplifying the heavenly spheres. Linking both shapes is the human form (Melchizedek, 1992). In drawing these two shapes, Leonardo carefully constructed the image in such a way that the perimeter of the square (about thirty-five cm at the scale in Figure 15) is essentially equal to the circumference of the circle. The idea of a circle superimposed on a square, also known as the "circle squared," is a frequently occurring form in ancient works. For a discussion of the significance and occurrence of this image the interested reader is referred to the work of Michell (Michell, 1972).

Figure 15. Leonardo da Vinci's Canon of Porportions

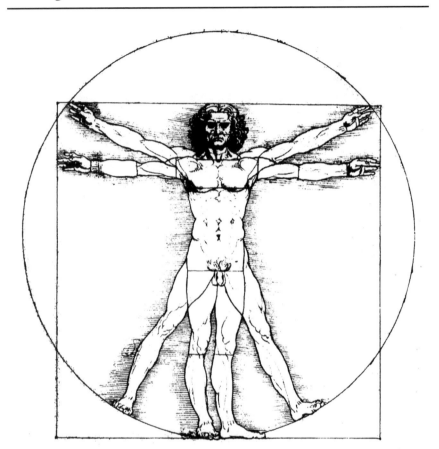

According to Leonardo, the study of anatomy or physiology is based not so much on measurement, but on perfect harmonies which are found in human proportions or ratios. It is very difficult to establish exact dimensions of the human figure which will remain the same among all types of individuals. However, the ratio or proportion between one part of the body and another appears to be less variable among different people and between the genders. Using this principle, we can begin to grasp the deeper significance of Leonardo's drawing. For example, Leonardo stated that, "The span of a man's outstretched arms is equal to his height" (MacCurdy, 1941, p. 206). This proportion is clearly depicted in Figure 15 where the height of the figure and length of its outstretched arms, each equal the length of one edge of the square. It is beyond the purpose of this book to provide an in-depth analysis of the numerous hidden proportions displayed in Leonardo's drawing. Rather, we refer the interested reader to the works of Doczi (Doczi, 1981, p. 93) and Lawlor (Lawlor, 1982, p. 59).

The Importance of the Location of the Navel and the Base of the Spine

Another key aspect of Leonardo's drawing is the location of the base of the spine and the navel in Figure 15. In modern physiology the base of the spine is an area at the bottom of the spinal cord called the sacrum (from the Latin *os sacrum*, translated as the "sacred bone") which is composed of the five sacral vertebrae fused together. This area corresponds to the body's center of gravity and is its geometrical center (Doczi, 1981). The location of the sacrum or base of the spine in Leonardo's drawing is positioned at the center of the square. As for the navel, it is situated at the mark on the abdomen where the umbilical cord was attached during gestation (pregnancy). In Leonardo's drawing this corresponds to the center of the circle. Another fascinating, but more complicated, way to locate the navel is to use a calculation based on the phi ratio[5] (Roger Silber, personal communication). The phi ratio, also known as the divine proportion or golden ratio, is derived from the golden mean rectangle by dividing the longer side by the shorter side. The ubiquity of this ratio in nature has aroused much interest regarding the importance and significance of this proportion. Interestingly, this ratio is found throughout Leonardo's drawing and can even be derived from it (for more information refer to Huntley, 1970; Doczi, 1981; Lawlor, 1982).

The distance between the base of the spine and the navel (about 0.95 cm on Leonardo's diagram in Figure 15) provides an important length which

---

[5] In order to calculate the location of the navel on the human figure in Leonardo's Canon of Proportions using the phi ratio (1.618...) the following formula is employed:

> (*height of the figure*) × (*1/phi ratio*) = *location of navel*
> 8.8 cm × (1/1.618) = 5.4 cm

The final number calculated is the distance from the base of the square to where the navel is situated.

Figure 16. Locating the Stat Tetrahedron in
Leonardo da Vinci's Canon of Porportions

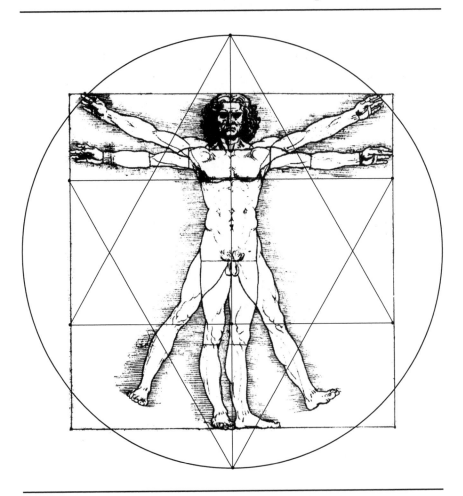

connects together the square and the circle. The diameter of the circle turns
out to be the length of one edge of the square plus twice the distance
between the navel and the base of the spine. Moving the circle down by the
distance between the base of the spine and the navel will superimpose the
circle symmetrically on top of both the human figure and the square which
have remained stationary (refer to Figure 16). In this revised diagram of
Leonardo's work the common center of both the square and the circle is
located at the base of the spine (the center of the circle is no longer at the
position of the navel because the circle has been moved down).

The Star Tetrahedron is Embedded in the Canon of Proportions

With the modified drawing in Figure 16 (the circle superimposed symmetrically on both the human form and the square) it is possible to locate a star tetrahedron surrounding the human image. The star tetrahedron is used in geometry to describe two interlocking tetrahedra. Together the two tetrahedra form an eight-pointed geometrical solid. In Sacred Geometry the two tetrahedra which constitute the star tetrahedron represent male and female energy fields. The base of the "male" tetrahedron crosses the knees and forms an apex above the head and extends into the space in front of us. The base of the "female" tetrahedron crosses just below the chest and forms a nadir below the feet and extends into the space behind us. Both energy fields coexist in the human form and together create the image of the star tetrahedron. In Figure 16 we present the proper construction of the star-tetrahedron surrounding Leonardo's image. Although it is beyond the scope of the current discussion to elaborate on this, it is important to realize that, in order to draw the star tetrahedron surrounding the human form, it must be placed in the proper orientation (Roger Silber, personal communication; Melchizedek, 1992; Frissell, 1994).

Leonardo's Canon of Proportions Displays the Constitution of the Universe

So far we have arrived at a basic understanding of the symmetry, proportions and key geometrical relationships concealed in Leonardo's Canon of Proportions. Using these principles, we will now locate in Leonardo's drawing the structure of the Constitution of the Universe.

Our first connection stems from the importance of the base of the spine. This point in the physiology, corresponding to the body's center of gravity, matches the creation of the first syllable of the Constitution of the Universe, "A-K." According to Rig Veda, "A" represents the state of all possibilities, which collapses to the point value of "K." Likewise, all the proportions and energy fields which surround the human physiology ("A") revolve around the base of the spine ("K").

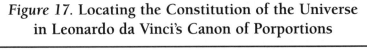

Figure 17. Locating the Constitution of the Universe
in Leonardo da Vinci's Canon of Porportions

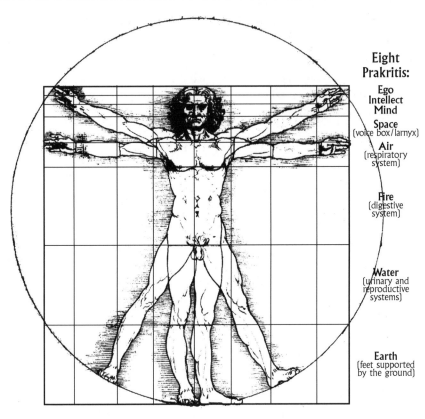

The next level in the Constitution of the Universe is the unfoldment of the
first syllable "A-K" into the eight syllables or prakritis of the first pada
(refer Figure 1). Likewise we have located a similar structure in Leonardo's
drawing. If we look back at the original drawing of Leonardo, reproduced
in Figure 15, we will notice that Leonardo drew a number of horizontal
and vertical lines on the human form. By extending only the horizontal
lines to the edge of the square we will obtain eight divisions or levels on
the diagram. These eight divisions match surprisingly well with the eight
prakritis. The following parallels equate the eight prakritis of the
Constitution of the Universe with the eight divisions on Leonardo's
drawing (refer to Figure 17 for summary):

- The first three divisions of Leonardo's Canon of Proportions divide the head into three levels: top of the head, forehead and the region between the eyes and nose. In Western science the brain and central nervous system have often been associated with the mental faculties and the information processing center. It is very intriguing that Leonardo assigned three of the eight levels in his drawing just to the head or brain region alone. Moreover, these three levels correspond to the first three prakritis: Ahamkara (translated as ego), Buddhi (translated as intellect) and Manas (translated as mind).

- The fourth division on Leonardo's work, the neck or throat region, is the site of the larynx (voice box) which constitutes the upper portion of the respiratory tract. One of the main functions of the larynx is to create speech and sound. The properties of sound and spatial movement of air associated with the larynx match the properties attributed to the fourth prakriti, Akasha (translated as space).

- The fifth part of Leonardo's drawing contains the chest region located between the base of the neck and the top of the abdominal region. One of the major organs of this system, the lungs, form a key part of the respiratory system and closely correspond to the fifth prakriti, Vayu (translated as air). Air or oxygen is transported into the body via the lungs and in exchange carbon dioxide is taken out.

- The sixth partition in the Canon of Proportions takes in the area of the body between the chest and lower abdomen. In this region the major components of the gastrointestinal tract responsible for digestion and metabolism matches the sixth prakriti, Agni (translated as fire). Fire is known to break down materials into smaller parts. This property has often been related to the body's "internal fire" of metabolism and digestion which breaks down food into easily assimilable particles.

- The seventh division in Leonardo's diagram consists of the body region extending from the lower abdomen down to the knees. Physiologically this area may be associated with the urinary and reproductive systems which match the seventh prakriti, Jala (translated as water). One way to understand water is to think of fluids or liquids. Both the urinary and reproductive systems have a number of fluids associated with them (e.g., urine, semen, menstrual discharge and amniotic fluid).

• The eighth and final level in Leonardo's drawing encompasses the feet and lower legs. Of all the components of the physiology these two are situated closest to the ground and hence correspond aptly with the eighth prakriti, Prithivi (translated as Earth). The Earth or ground is not only in constant contact with the soles of our feet, but it also supports our entire structure.

From the basic eight syllables of the first pada of Rig Veda, the next level of expansion in the Constitution of the Universe is the interpretation of each of these eight syllables with respect to rishi (knower), devata (process of knowing) and chhandas (known). This produces a total of twenty-four units. Similarly, Leonardo's Canon of Proportions embodies an arrangement of three. These three aspects are represented by the circle, the human form and the square, corresponding precisely with the three aspects of rishi, devata and chhandas, respectively. To the Ancients the circle represented heaven (the heavenly spheres) which is equated with the abstract nature of rishi. The square, however, is the symbol for the earth (firmness) which is well matched with the concrete and material nature of chhandas. Finally, the human form is that which connects or unites heaven and earth. This is precisely the nature of devata which provides a link between rishi and chhandas. All three aspects (the circle, the square and the human form) are superimposed on one another, thus locating a geometrical relationship in any one of these would also apply to the other two. Hence, the eight divisions or levels in Leonardo's drawing (described above), when viewed with respect to the human form, the circle and the square, will make up a total of twenty-four units corresponding to the twenty-four units in the Constitution of the Universe.

In the final stage of the Constitution of the Universe the eight syllables of the first pada of Rig Veda expand into sixty-four units. A similar structure was also located in Leonardo's Canon of Proportions by Melchizedek (Melchizedek, 1992; Frissell, 1994). As described earlier, there are a number of horizontal and vertical lines on Leonardo's original drawing. The horizontal lines we already extended in order to create the eightfold division corresponding to the eight prakritis. Now, if we simply extend the vertical lines to the edge of the square we will obtain a grid of sixty-four squares (refer to Figure 17). This 8 × 8 grid represents the energy centers around the body. The construction of these sixty-four squares from the eight divisions corresponds precisely with the elaboration of the sixty-four syllables from the eight prakritis in the Constitution of the Universe. Moreover, each of these sixty-four squares can be viewed with respect to the circle, the square and the human form, thereby creating a total of 192 units. This provides a complete match with the final level of the Constitution of the Universe where each of the sixty-four syllables of Rig Veda are interpreted with respect to rishi, devata and chhandas (a total of 192 units).

In light of these intriguing correlations, it seems likely that Leonardo knew about the fundamental structure of creation, which we call the Constitution of the Universe, when he drew the Canon of Proportions. Just by looking at this image, the structure of the Constitution of the Universe template jumps out. The first aspect that stands out is the three-fold nature of the circle, the human form, and the square (corresponding to rishi, devata and chhandas). Next, by simply extending the lines Leonardo drew on his image we obtain eight distinct horizontal levels and a grid of sixty-four squares. In this drawing Leonardo expressed his deep conviction that through the image of our own body anything and every-thing can be known. To know yourself is to know it all. He believed that "every part is disposed to unite with the whole, that it may thereby escape from its incompleteness" (Doczi, 1981, p. 95). The fact that Leonardo incorporated the same structure as that of the Constitution of the Universe into his Canon of Proportions and other works may explain why his accomplishments enjoy undying fame. We conclude this section with a quote from Leonardo expressing his respect for the Natural world:

> *"Though human ingenuity may make various inventions which, by the help of various machines, answer the same end, it will never devise invention more beautiful, nor more simple, nor more to the purpose, than nature does; because in her inventions nothing is wanting and nothing is superfluous, and she needs no counterpoise when she makes limbs proper for motion in the bodies of animals."*

– **Leonardo da Vinci**
Reynal and Company (1967). *Leonardo da Vinci.*
Japan: Dainippon Printing Co., Ltd., p. 30.

SECTION II

MAGIC SQUARES AND CUBES

*"The peculiar interest in magic squares …
lies in the fact that they possess the charm of
mystery. They appear to betray some hidden
intelligence which by a preconceived plan
produces the impression of intentional
design, a phenomenon which finds its close
analogue in nature … Magic squares are
conspicuous instances of the intrinsic
harmony of number, and so they will serve as
an interpreter of the cosmic order that
dominates all existence. Though they are a
mere intellectual play they not only illustrate
the nature of mathematics, but also,
incidentally, the nature of existence
dominated by mathematical regularity."*

W.S. Andrews, 1960, p. vii

Since ancient times the construction of magic squares has been an area of mysterious charm and great entertainment. We find evidence for their existence in China, Babylon, India, Arabia, Jaina, Egypt, Greece and indeed throughout the world. The name "magic squares" originated in the Eastern Mediterranean where they were used in the Ancient Egyptian mystery schools and by the Pythagoreans who believed that all things were regulated by the harmony of numbers. Although there is no known practical use for the study and construction of magic squares, they have influenced the thinking of many individuals throughout the ages. Construction of these figures is not like solving simple puzzles, but rather requires the ability to observe a balanced relationship between numbers. Our purpose in this section is to describe how to construct magic squares and cubes and to demonstrate that specific classes of these magic figures relate to the numerical structure of the Constitution of the Universe.

Description of Magic Squares

A magic square is a term used to describe an aggregation of numbers arranged in the configuration of a square. These numbers, however, must be arranged in such a way that the summation of all the numbers in each row, each column and the two corner diagonals equals the same amount. To construct a simple magic square, we first draw a square and then subdivide it into a certain number of identical subcompartments or cells. Then within each cell one number is placed. Construction of a magic square can begin in any cell desired.

Magic Squares are classified into two general categories. "Odd" magic squares are defined as a magic square where the total number of cells along one edge of the square is equal to an odd number (i.e., 3, 5, 7, 9, etc.). "Even" magic squares are defined simply as a magic square where the total number of cells along one edge of the square is equal to an even number (i.e., 4, 6, 8, 10, etc.). A 3 × 3 square is the smallest set of numbers capable of forming a magic square arrangement. There however, is no limit to how large a magic square may be constructed.

In addition to the two general categories of magic squares, there are four fundamental characteristics which help us to comprehend the virtually infinite variety of possible magic squares. These four basic aspects are as follows: (n) the number of cells which occupy one edge of the square; (S) the summation of any row, column or corner diagonal; (A) the lowest or starting number in the square; and, (β) the increment between the numbers. These last two characteristics may be somewhat difficult to comprehend at first. For clarity we will provide an example. To construct a magic square the first step is to write down a number, which for sake of argument will be the lowest number in the magic square. Usually this number is "1," but in reality any value can be used (e.g., -8, 4, etc.). Once we have established an initial number (A), we then proceed to write down the other numbers (each number higher than the previous). We need, however, to decide what will be the increment (β) between the numbers. Typical magic squares have an increment of one (e.g., 1, 2, 3, 4, etc.), but we could just as easily construct a magic square with an increment of three (e.g., 1, 4, 7, 10, 13, etc.). The increment (β) must however remain equal for any one given magic square.

Mathematically these four characteristics are related by the following equation (Andrews, 1960):

$$S = A\,n + \beta\,n/2\,(n^2 - 1)$$

Where:

if $A = 1$ and $\beta = 1$,

then $S = n/2\,(n^2 + 1)$.

Hence, by knowing any three of the characteristics we can determine the fourth. For example, if a magic square has eight cells on one side (n), the initial number equal to one (A) and the increment also equal to one (β), then the summations are equal to 260. This means that each row, column and corner diagonal in a magic square with these characteristics will add up to 260.

Now we will provide some actual examples of magic squares and describe a few methods for constructing them. It is important to realize the magic squares should not be constructed by the method of "trial and error" or "try and try again." Magic squares are not simple puzzles, but are expressions of fixed laws of geometrical order. It is, however, important to note that no general rule for the construction of magic squares has been established. Although numerous methods for constructing them exist, we will discuss only a few basic ones to give the reader an idea of how to make them. New methods and original magic squares continue to be developed providing a fruitful area of research and recreation.

For the purpose of this section our prime focus will be on the 8 × 8 magic squares. Although we do realize that numerous other types of magic squares exist that are important and interesting, it is beyond the scope of this book to provide an in-depth discussion of the entire area of magic squares. Rather, we have limited our attention to the 8 × 8 magic squares for several reasons. First, the 8 × 8 magic squares have been the focus of several famous individuals and are found in a number of ancient cultures. Second, this type of magic square has an intriguing relationship to the game of chess which will be described shortly. Last, the 8 × 8 magic square is a precise reflection of the numerical structure of the Constitution of the Universe.

Construction of the 3 × 3 Magic Square

Prior to examining 8 × 8 magic squares, we wish to ensure that the reader is familiar with the basic concepts of a magic square. In order to do this, we have provided two examples of the simplest type, 3 × 3 magic square (refer to Figure 18). Both these examples are derived from the work of Andrews (Andrews, 1960).

In the first example both the starting number (A) and the increment (β) are equal to one. Because it is a 3 × 3 magic square, the number of cells per side (n) is equal to three. Thus, using the basic equation described above, the summation (S) of any row, column or corner diagonal is equal to 15. The calculations for each row, column and corner diagonal are tabulated in Figure 18. The second example follows from the first. Two variables have been changed, so that both the starting number (A) and the increment (β) are equal to two instead of one. This means that the summation (S) is 30. Again, the calculations for the summation of each row, column and corner diagonal are summarized in Figure 18.

Construction of the 8 × 8 Magic Square

Although the method of "trial and error" may be applicable to solving simple magic squares, like the 3 × 3 arrangement, when we proceed to higher order magic squares like the 8 × 8 magic squares, this is no longer the case. Higher order magic squares should be constructed using geometrical patterns or methods. For the 8 × 8 magic square various methods have been developed. We have presented one of these methods in Figure 19. The diagrammatic images appearing to the left of the example of the 8 × 8 magic square provide the method for its construction (Andrews, 1960). As can be seen, such diagrammatic images are relatively easy to construct. The procedure for translating this diagrammatic image into the 8 × 8 magic square is as follows. We start by writing the first row or level of the magic square. The eight numbers for this first row are obtained from the first diagrammatic image by starting with the number 1 (top left) and tracing or following the line down to the number 8 (i.e., 1 → 63 → 62 → 4 → 5 → 59 → 58 → 8). These eight numbers constitute the first row of the magic square in Figure 19. The second row is constructed from the second diagrammatic image starting with the number 56 (top right) and ending with the number 49. Similarly the third and fourth rows of the

Figure 18. Two Examples of the 3 × 3 Magic Square

3 × 3 Magic Square:

- Starting Number (A) = 1
- Increment (β) = 1
- Summation (S) = 15
- # of Cells/Side (n) = 3

8	1	6
3	5	7
4	9	2

Rows:

8 + 1 + 6 = 15
3 + 5 + 7 = 15
4 + 9 + 2 = 15

Columns:

8 + 3 + 4 = 15
1 + 5 + 9 = 15
6 + 7 + 2 = 15

Corner Diagonals:

8 + 5 + 2 = 15
4 + 5 + 6 = 15

3 × 3 Magic Square:

- Starting Number (A) = 2
- Increment (β) = 2
- Summation (S) = 30
- # of Cells/Side (n) = 3

16	2	12
6	10	14
8	18	4

Rows:

16 + 2 + 12 = 30
6 + 10 + 14 = 30
8 + 18 + 4 = 30

Columns:

16 + 6 + 8 = 30
2 + 10 + 18 = 30
12 + 14 + 4 = 30

Corner Diagonals:

16 + 10 + 4 = 30
8 + 10 + 12 = 30

Adapted from Andrews, W.S. (1960). *Magic Squares and Cubes.*
New York: Dover Publications, Inc., pp. 54 - 55.

magic square are constructed using the third (begin top right) and fourth (begin top left) diagrammatic images, respectively. To construct the fifth row of the 8 × 8 magic square, we need to follow the lines of the diagrammatic images in the reverse direction. This means we start with the fourth image on number 33 (begin bottom right) and proceed backwards to the number 40. These eight numbers give us the fifth row of the 8 × 8 magic

Figure 19. An Example of an 8 × 8 Magic Square

1	63	62	4	5	59	58	8
56	10	11	53	52	14	15	49
48	18	19	45	44	22	23	41
25	39	38	28	29	35	34	32
33	31	30	36	37	27	26	40
24	42	43	21	20	46	47	17
16	50	51	13	12	54	55	9
57	7	6	60	61	3	2	64

Totals of Rows, Columns
and Corner Diagonals = 260

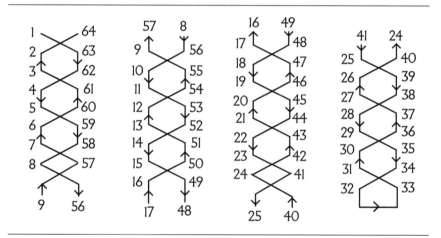

Adapted from Andrews, W.S. (1960). *Magic Squares and Cubes.*
New York: Dover Publications, Inc., p. 25.

square. Likewise, rows six, seven and eight are constructed from the third (begin bottom left), second (begin bottom left) and first (begin bottom right) diagrammatic images, respectively. This completes the construction of this particular 8 × 8 magic square where the summation of all numbers for each row, column and corner diagonals is equal to 260.

Numerous other methods for the construction of 8 × 8 magic squares are possible. For example, the lines in diagrammatic images used in the above example can be modified into various intricate patterns which will also produce magic squares. Further discussion of such methods would be too involved for the present discussion. A good review is provided by Andrews (Andrews, 1960).

Franklin Magic Squares

Our next example of 8 × 8 magic squares is based on the work of one of the Founding Fathers of the United States, Benjamin Franklin. The magic squares constructed by Benjamin Franklin, called "Franklin Magic Squares," are of two types: the 8 × 8 and the 16 × 16. These two Franklin Magic Squares have attracted particular attention because of their unique and astonishing properties. In this section we will focus on the 8 × 8 Franklin Magic Square reproduced in Figure 20 (Andrews, 1960). Apart from having the normal properties of a magic square (i.e., each row, each column and the two corner diagonals adding up to the same amount), this 8 × 8 Franklin Magic Square has a number of other remarkable character-istics. For example, in addition to the eight rows and eight columns which add up to 260, there are thirty-two other "bent diagonals" which also sum up to 260. These include sixteen horizontal bent diagonals and sixteen vertical bent diagonals. Each "bent diagonal" is composed of eight cells with four of the cells ascending diagonally and the other four cells descending diagonally. To understand a horizontal bent diagonal we will locate an example using Figure 20: start on the number 11 and descend diagonally to the number 15 (i.e., 11 → 58 → 57 → 15), and then proceed to the number 18 and ascend to the number 22 (i.e., 18 → 40 → 39 → 22). Vertical bent diagonals are constructed in a similar way (they are horizontal bent diagonals rotated 90 degrees). Another interesting property of the 8 × 8 Franklin square is that the sum of numbers contained in any 2 × 2 subsquare will equal 130. Moreover, the sum of any four numbers that are arranged diametrically equidistant from the center of the Franklin square is also equal to 130.

A summary of the special properties of the Franklin 8 × 8 Magic Square is displayed in the configuration of cells located next to this square in Figure 20 (Andrews, 1960). Each configuration of cells represents a different property. Any place where these particular configurations can be located in the Franklin square, the sum of numbers enclosed by those cells will always equal the amount indicated (i.e., 130 or 260). How Benjamin

Figure 20. The Franklin 8 × 8 Magic Square

52	61	4	13	20	29	36	45
14	3	62	51	46	35	30	19
53	60	5	12	21	28	37	44
11	6	59	54	43	38	27	22
55	58	7	10	23	26	39	42
9	8	57	56	41	40	25	24
50	63	2	15	18	31	34	47
16	1	64	49	48	33	32	17

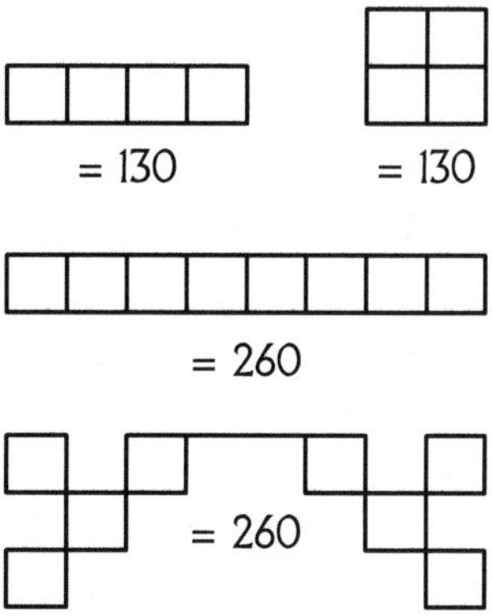

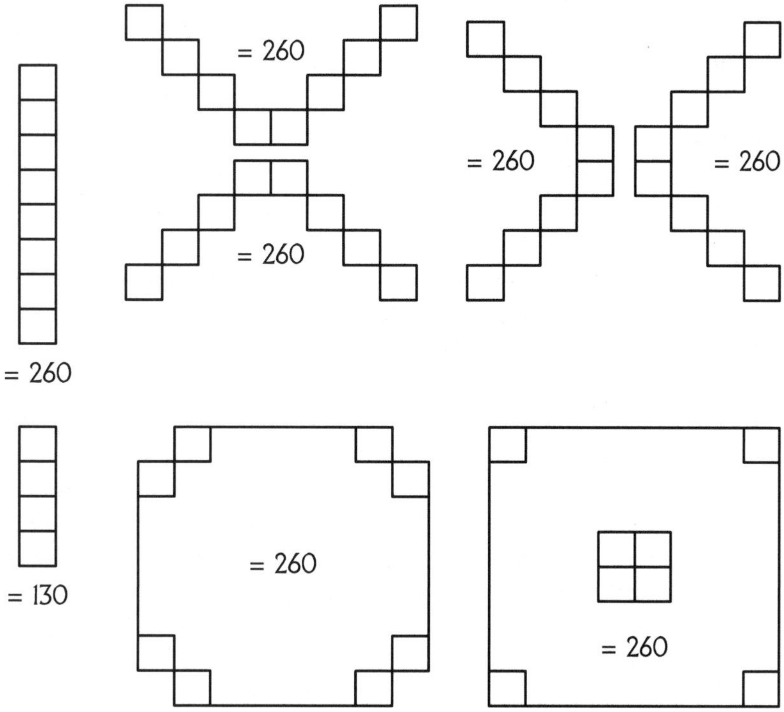

Adapted from Andrews, W.S. (1960). ***Magic Squares and Cubes.***
New York: Dover Publications, Inc., pp. 96 - 97.

Franklin constructed his 8 × 8 magic square is unknown; all we have is the final work. Andrews (Andrews, 1960), however, did describe a tedious process by which this square could be built, but it is probably not the method employed by Franklin.

Indian Magic Squares

Another fascinating example of 8 × 8 magic squares is found in the "Jaina" or Indian magic squares. Since time immemorial the Brahmins of India have been known as great experts in the formation of magic squares. Often patterns of these magic squares appear in Indian art and are worn by the natives as amulets. Indian magic squares, in addition to having the same properties as other magic squares, have several other characteristics. One of these properties, as already demonstrated in the Franklin 8 × 8 magic square, is that any cluster of four cells will sum up to 130. Another intriguing characteristic is that these squares will retain the magic square properties even when one or more of the columns or rows are interchanged.

Description and Examples of Magic Cubes

In addition to magic squares, other magic geometrical figures have been constructed. One example worth mentioning is the magic cubes. These are a little more challenging to construct than the magic squares because the interactions between the various numbers are more complex. By definition a magic cube is an aggregation of numbers arranged in cubic form such that the summation of all numbers in each row or column which runs parallel to any of the edges and the four corner diagonals equals the same amount (note that diagonals go **through** the cube, not across outside square faces). For the smallest even magic cube, the 4 × 4 × 4, there are fifty-two columns adding up to 130: sixteen vertical columns, sixteen horizontal (front and back), sixteen horizontal (left to right), and four diagonal rows uniting the four pairs of opposite corners (Andrews, 1960).

Magic cubes have the same four characteristics embodied by magic squares. As described earlier, these are as follows: (n) the number of cells occupying one edge of the cube; (S) the summation of any row, column or corner diagonal (through the cube); (A) the lowest or starting number

Figure 21. Construction of a 4 × 4 × 4 Magic Cube

Part I: An Example of a 4 × 4 × 4 Magic Cube with Diagramatic Image Showing Its Construction

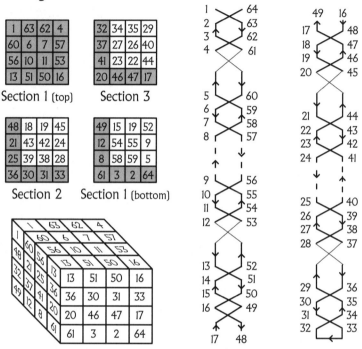

Section 1 (top)   Section 3

Section 2   Section 1 (bottom)

Note: One example of diagonal calculation would be as follows: 1 + 43 + 22 + 64 = 130

Part II: Another Way to View the Construction of the 4 × 4 × 4 Magic Cube: Each of Four Sections with Numbers 1 to 64 Written in Arithmetical Order

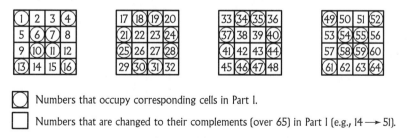

⬭ Numbers that occupy corresponding cells in Part I.

▢ Numbers that are changed to their complements (over 65) in Part I (e.g., 14 ⟶ 51).

Adapted from Andrews, W.S. (1960). *Magic Squares and Cubes.*
New York: Dover Publications, Inc., pp. 78 - 79.

in the cube; and, (β) the increment between the numbers. Because we are dealing with cubes rather than squares, the mathematical formula which relates these four basic aspects is slightly different:

$$S = A\,n + \beta\,n/2\,(n^3 - 1)$$

Where:

if $A = 1$ and $\beta = 1$,

then $S = n/2\,(n^3 + 1)$.

The difference between these two formulas for magic cubes and those for magic squares is that the value for "n" inside the parentheses is cubed instead of squared.

As with magic squares, there are numerous examples of magic cubes and methods for their construction. In Figure 21 we have provided a simple example of a $4 \times 4 \times 4$ magic cube and diagrammatic images which show how it is constructed. The diagrammatic images to the right of the magic cube are translated into the numbers on the sections of the magic cube similar to the method described above for the diagrammatic images for the 8×8 magic square in Figure 20. That is, for section 1 (top), the first row is obtained by starting on the number 1 in the diagrammatic image and tracing the line to number 4 ($1 \rightarrow 63 \rightarrow 62 \rightarrow 4$). Continuing in this fashion, the second line is similarly constructed by starting on the number 60 and ending on the number 57 ($60 \rightarrow 6 \rightarrow 7 \rightarrow 57$). Thus the entire $4 \times 4 \times 4$ magic cube is systematically and easily constructed. Another way to view the construction of this magic cube, without the diagrammatic image, is presented in Part II of Figure 21. The first step is to write down in each of the four sections of the $4 \times 4 \times 4$ magic cube the numbers 1 to 64 in arithmetical order (refer to Figure 21). Next, all the numbers which have circles placed around them remain the same for the final magic cube. All other numbers, however, must be changed to their complementary number. To calculate complementary numbers for this magic cube use the following formula:

65 - number = complement of the number

For example the complement of the number 8 is the number 57 ($65 - 8 = 57$). The general formula for any magic cube is $(n^3 + 1)$ - number, where n is the number of cells along one edge of the cube. Once all the numbers are changed to their complementary numbers, this completes the construction of the magic cube in Figure 21.

If we want to construct an 8 × 8 × 8 magic cube, it is simply an expansion of the 4 × 4 × 4 magic cube, just as the 8 × 8 magic square is an expansion of the 4 × 4 magic square. The same diagrammatic images used to make the 4 × 4 × 4 magic cube can be modified to construct the 8 × 8 × 8 magic cube. As described above, the 4 × 4 × 4 magic cube has fifty-two columns which add up to the same amount. In the 8 × 8 × 8 magic cube there are 192 columns which sum up to give the same amount. This number turns out to be equal to 2,052.

Magic Squares and Cubes Resemble the Structure of the Constitution of the Universe

Having arrived at a basic understanding of magic squares and figures, we will now locate in the 8 × 8 magic squares and the 4 × 4 × 4 and 8 × 8 × 8 magic cubes the numerical structure of the Constitution of the Universe. Our first connection relates to the first level of unfoldment in the Constitution of the Universe, from the first syllable "A-K" into the eight syllables or prakritis of the first pada (refer to Figure 1). By definition, an 8 × 8 magic square must have the eight cells in each row, each column and the two corner diagonals equal to the same amount. We propose that these groups of eight cells, on which the magic square is based and defined, correspond to the eight fundamental aspects of the Constitution of the Universe. Likewise, the 8 × 8 × 8 magic cube is also based on groups of eight cells which all add up to the same number. It is interesting to note that the summation number for each row, each column and the two corner diagonals in an 8 × 8 square is equal to 260 (provided that the increment [β] and the starting number [A] are equal to one). (As will be discussed in Chapter 8, Section VII, 260 is also the number of days in the Tzolkin or Sacred Mayan Calendar which also has hidden in its structure the Constitution of the Universe).

From the eight basic syllables of the first pada of Rig Veda, the next level of expansion in the Constitution of the Universe is the interpretation of each of these eight syllables with respect to rishi (knower), devata (process of knowing) and chhandas (known). This produces a total of twenty-four units. Similarly for any magic square or magic cube there are four fundamental characteristics – three of these correspond to rishi, devata and chhandas. The first characteristic, the starting number (A) or the lowest number in the magic figure, matches the nature of rishi which represents the first impulse of intelligence prior to the creation of devata and chhandas. The second characteristic in a magic figure, the increment or increasing number (β), determines the numerical interval between two successive numbers. This matches the qualities embodied by the devata interpretation which connects rishi to chhandas. The third aspect, the summation value (S) which is calculated after the magic figure is constructed, is equated to the expressed or material nature of the chhandas interpretation. These three characteristics of a magic square or cube when viewed with respect to each of the eight basic cells (which sum up to the same amount) comprise a total of twenty-four units corresponding to the twenty-four units of the Constitution of the Universe. We propose that the fourth characteristic in a magic figure, the number of cells along one edge of the square (n), embodies what is known in the Rig Veda as the samhita value or the togetherness of rishi, devata and chhandas. For a magic square or cube the number of cells along one edge defines the entire matrix.

In the next stage of elaboration of the Constitution of the Universe the eight syllables of the first pada of Rig Veda expand into sixty-four units. Similarly, all 8 × 8 magic squares contain a total of exactly sixty-four subsquares or numbers where the clusters of eight cells (rows, columns and corner diagonals) reside. Sixty-four is also the total number of subcubes or numerals in the 4 × 4 × 4 magic cube.

The final correspondence with the Constitution of the Universe stems from the sixty-four basic syllables of Rig Veda interpreted with respect to rishi, devata and chhandas, making a total of 192 units. Likewise the number in each of the sixty-four subsquares of an 8 × 8 magic square is determined with respect to three values (a total of 192 units): the starting number (A), the increment number (β) and the value of the summation (S). These three values also pertain to any subcube in the 4 × 4 × 4 or 8 × 8 × 8 magic cubes. It is interesting to note that the total number of columns which sum up to the same amount in an 8 × 8 × 8 magic cube is 192, this being the number of syllables contained in the elaborated aspect of the Constitution of the Universe (sixty-four syllables times the three interpretations of rishi, devata and chhandas). In view of the correspondences just described, it is significant and interesting that the 8 × 8 magic squares and cubes have been given special attention in ancient cultures and more recently by famous people.

Figure 22. The 8 × 8 Gameboard of Chess:
Possible Movements of the Knight and the "Knight's Tour"

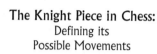

The Knight Piece in Chess:	The "Knight's Tour":
Defining its	One Example of the
Possible Movements	Numerous Possibilities

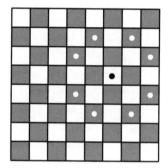

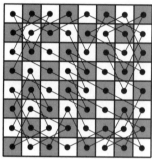

● Knight

○ 8 Possible Moves
  of the Knight

Adapted from
Falkener (1961). p. 350.

The 8 × 8 Magic Square is Linked to the Game of Chess

In concluding this section we would like to describe a fascinating link between the 8 × 8 magic square and the game of chess. To do this we need to describe what is known as the "Knight's Tour." Of all the pieces in the game of chess the Knight piece makes the most unique move. It can roughly be described as an L-shaped jump to a square of the opposite color. Figure 22 graphically displays the possible movements of the Knight from any given square. If the Knight piece is not located too near the edge of the game board, it has a total of eight possible movements. In other words, each move of the Knight consists of two parts: first, it can jump two squares forward, backwards, or to either side, and then one square to the left or right.

The "Knight's Tour," which is perhaps as old as the game of chess itself, is an intriguing way to cover all sixty-four squares on the chess board in an array of continuous Knight moves, without landing on the same square twice. Although there are many possible solutions to complete the Knight's Tour, there are two general categories. First, the Knight's Tour can be completed in such a way that the Knight does not finish the "tour" on the same square from which the "tour" was started. The second way, also called the "perfect" solution to the Knight's Tour, is completed when the Knight finishes the "tour" on the same square from which the "tour" began.

In Figure 22 we have given one example of a "perfect solution" to the Knight's Tour. As is demonstrated in this example, if the path of the Knight's Tour is drawn out, it often forms an orderly pattern. Moreover it is possible to construct a magic square out of certain solutions to the Knight's Tour (Falkner, 1961). Such solutions to the Knight's Tour can begin at any square on the board. The procedure for translating the moves of the Knight into the series of numbers used to construct the magic square is as follows. First, label all the squares on the board from one to sixty-four in arithmetical order. Then the series of numbers for the magic square are generated depending on what square the Knight passes over during its "tour." For the interested reader it is also fascinating to note that it is possible to complete the Knight's Tour in such a way to resolve the chess board in four quarters of sixteen cells each, where each quarter is a magic square as well as the entire 8 × 8 matrix.

The history and widespread occurrence of chess in most ancient civilizations, along with its interesting 8 × 8 squared game board has a deep connection with the Constitution of the Universe. However, instead of going into detailed discussion of this topic here, we refer the reader to Chapter 8, Sections I and II. Described in these sections are the ancient Sumerian and ancient Persian belief that the secret to understanding the creation of the universe is displayed in the game of chess.

CHAPTER 8:

ANCIENT CIVILIZATIONS AND THE CONSTITUTION OF THE UNIVERSE

SECTION I

ANCIENT MESOPOTAMIA – SUMERIAN COSMOGONY

The word Mesopotamia (meaning "between the two rivers") is a term used to describe an area of the ancient world corresponding to modern-day Iraq, South-East Turkey and Eastern Syria, an area where a number of major civilizations were born. This area may have become a rich haven for culture because it is located between the Tigris and Euphrates rivers. Fertile lands extending between these rivers could have easily supported a large population. One of the earliest known civilizations to emerge from the banks of these rivers was the ancient Sumerians.

Archeological evidence from Sumerian hieroglyphs, tiles and stones date back to between eight and ten thousand years ago. Although there are few written Sumerian texts existing from this far back, we have knowledge of the symbols and other such codes they left behind. Often such relics conceal wisdom or understanding of our place in creation and how the universe originated. Based on this premise Edmond Szekely was able to decipher, from the ancient Sumerian hieroglyphs and symbols, an integrated understanding of Sumerian cosmogony, or the study of the origin of the universe. Szekely, an author of more than eighty books, spoke ten modern languages. In this section we present evidence, primarily based on the work of Edmond Szekely (Szekely, 1994), that the understanding of Sumerian cosmogony is identical with the numerical structure of the Constitution of the Universe. We suggest that the reader study this section together with the next section on ancient Persia, for two reasons. First, both these ancient cultures had similar understandings of the origin of creation (it is possible that ancient Persia derived its wisdom from Sumeria, Babylon and Assyria). Second, the information presented in both these chapters is based on the work of Szekely.

The Origin of the Universe

According to ancient Sumerians the universe is governed by two basic forces, human consciousness and nature – together they provided the foundation for their code of life. Human consciousness provided inner guidance while mother nature provided outer guidance (Szekely, 1994). Their understanding of the origin and structure of the universe was based on the simple principles of geometry. According to Sumerian Cosmogony (study of the origin of the universe), the Creator constructed the universe in sixteen consecutive motions.

There was neither motion nor movement at the origin of the universe. This origin or beginning was symbolized by a point or a dot which represented the Creator (refer to Figure 23). All combinations and permutations of everything in the universe emerged from this point. The following passage from Szekely's work aptly summarizes the properties of the origin of the universe as symbolized by this point:

> "It [the point] was the symbol of the beginning, out of which everything came. It had no dimensions, no materiality, and existed neither in time nor in space, neither in force nor in matter. It signified the basic power of the universe, the creative principle which weaves the universe – with time and space, force and matter manifesting as the creation. Not until the point moves is anything created in the universe."

> – Edmond Bordeaux Szekely
> (1994) *Archeosophy, A New Science*
> USA: International Biogenic Society, pp. 25 - 26

Four Primal Movements
Construct the Cosmic Nubulae

As described in the above excerpt, nothing is created until there is movement or motion. For the Sumerians the first movement of the creative principle was the construction of a vertical line directly above the

Figure 23. Sumerian Cosmogony I:
The Origin of the Universe

Stage One:
From a Point to the Cosmic Nebulae
(Genesis of Time, Space, Matter & Force)

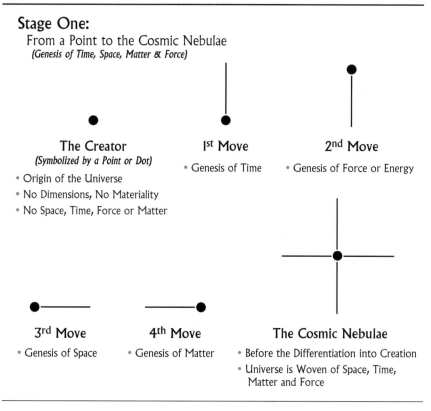

The Creator
(Symbolized by a Point or Dot)

* Origin of the Universe
* No Dimensions, No Materiality
* No Space, Time, Force or Matter

1st Move

* Genesis of Time

2nd Move

* Genesis of Force or Energy

3rd Move

* Genesis of Space

4th Move

* Genesis of Matter

The Cosmic Nebulae

* Before the Differentiation into Creation
* Universe is Woven of Space, Time, Matter and Force

Adapted from Szekely, E.B. (1994). *Archeosophy, A New Science.*
U.S.: International Biogenic Society.

point of origin (refer to Figure 23). The symbol of a point with a vertical line above it was the Sumerian symbol for the genesis of Time. The second movement of the creative principle is the drawing of a vertical line directly below the point of origin (refer to Figure 23). This motion symbolized the creation of Force or Energy. Together the first and second motions are represented by two lines, one above and the other below the point. In ancient Sumerian pictographs this frequently appearing image represents Time and Force or the movement of energy through time.

The third motion of the creative principle, represented by a horizontal line extending to the right of the dot (refer to Figure 23), symbolizes the creation of Space. After this we have the fourth motion, represented by a line in the opposite direction, a horizontal line stretching to the left of the point, symbolizes the genesis of Matter (refer to Figure 23). As with Time and Force, the Sumerians often symbolized Space and Matter together in one symbol which portrayed the concept of volume or mass.

Together these first four movements of the creative principle construct what is known as the "Cosmic Nebulae" (refer to Figure 23). In this symbol the dot in the center represents the Creator and the four lines radiating outward correspond to the four primal entities, Time, Force, Space and Matter. According to the ancient Sumerians, the Cosmic Nebulae signifies a kernel of unity and absolute potentiality containing all the information necessary to construct the diversity of the universe.

Sumerian Cosmogony and the First Unit of the Constitution of the Universe

From this initial description of Sumerian cosmogony we can draw our first correspondence with the Constitution of the Universe. The first unit of the Constitution of the Universe is represented in the Rig Veda by the letter "A" which signifies a state of all possibilities, unity and the total potential of natural law. This concept matches the Sumerian view of how the universe was created. According to Sumerian cosmogony, the origin of the universe began with the Creator or creative principle (corresponding to "A") which represents a state of unity, no dimensions, no space-time and a field of all possible combinations and permutations. The second unit of the Constitution of the Universe is represented in the Rig Veda by the letter "K" signifying the point value from which diversity is created. Similarly, the ancient Sumerians believed that the Cosmic Nebulae was a kernel of potentiality responsible for the diversification of the creative principle.

The Cosmic Nebulae Undergoes an Eight-Fold Elaboration

The next stage in the Sumerian cosmogony is an eight-fold elaboration of the four entities contained in the Cosmic Nebulae: Time, Force, Space and Matter. The elaboration of each entity occurs in eight precise moves.

Interestingly the number eight was considered the most important numeral in ancient Sumerian culture (Szekely, 1994). For this reason, the number eight is often found throughout Sumerian symbols as well as other cultures of this region (e.g., the ancient Persians).

The first eight-fold elaboration is the expansion of the entity Time (symbolized by the top vertical line on the Cosmic Nebulae). This elaboration is divided into two stages corresponding to the fifth and sixth movements of the Creator. In the first stage (fifth move), four new lines are drawn to the right of the symbol for Time. Similarly, in the second stage (sixth move), four new lines are drawn to the left of the symbol for Time. This eight-fold elaboration of Time symbolizes the creation of the eight seasons: spring, spring-summer, summer, summer-fall, fall, fall-winter, winter and winter-spring (refer to Figure 24).

The second eight-fold elaboration occurs with respect to the entity of Force (symbolized by the lower vertical line on the Cosmic Nebulae). This eight-fold elaboration or division evolves differently from that described for the entity of Time. The first step (corresponding to the seventh move) is the elaboration of Force into four categories of energy: the stars, the sun, the earth and humans. Each of these four energies is then further divided into good and evil to create a total of eight different types of energies (this represents the eighth move of the Creator). In this second eight-fold elaboration we have introduced a new concept, the idea of duality (refer to Figure 24).

The third eight-fold elaboration is the expansion of the entity Space (symbolized by the right horizontal line on the Cosmic Nebulae). As with the eight-fold division of Time, this elaboration is divided into two stages corresponding to the ninth and tenth movements of the Creator. The first stage involves the construction of four lines drawn parallel to and above the Space line (ninth move of the Creator). For the second stage, an additional four lines are constructed parallel to and below the Space line (tenth move). The construction of these eight new lines from Space correspond to the genesis of the eight cardinal points: east, southeast, south, southwest, west, northwest, north, and northeast (refer to Figure 24).

The fourth eight-fold elaboration occurs with respect to the entity of Matter (symbolized by the left horizontal line on the Cosmic Nebulae). This elaboration is divided into two steps corresponding to the eleventh and twelfth moves of the Creator. The first step (eleventh move) is the expansion of Matter into four categories: air, water, earth and fire. Each of

Figure 24. Sumerian Cosmogony II: The Origin of the Universe

Stage Two:
Eight-Fold Elaboration of Time, Force, Space and Matter

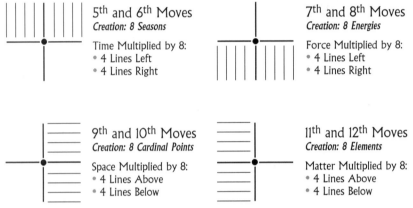

5th and 6th Moves
Creation: 8 Seasons

Time Multiplied by 8:
• 4 Lines Left
• 4 Lines Right

7th and 8th Moves
Creation: 8 Energies

Force Multiplied by 8:
• 4 Lines Left
• 4 Lines Right

9th and 10th Moves
Creation: 8 Cardinal Points

Space Multiplied by 8:
• 4 Lines Above
• 4 Lines Below

11th and 12th Moves
Creation: 8 Elements

Matter Multiplied by 8:
• 4 Lines Above
• 4 Lines Below

Stage Three:
Combination of All Entities into the Cosmic Tapestry of Sixty-Four Squares

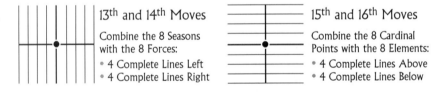

13th and 14th Moves

Combine the 8 Seasons
with the 8 Forces:
• 4 Complete Lines Left
• 4 Complete Lines Right

15th and 16th Moves

Combine the 8 Cardinal
Points with the 8 Elements:
• 4 Complete Lines Above
• 4 Complete Lines Below

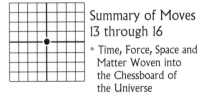

Summary of Moves 13 through 16
• Time, Force, Space and
  Matter Woven into
  the Chessboard of
  the Universe

Final Tapestry of the Universe
• 64 Squares
• 32 White Squares *(Good)*
• 32 Black Squares *(Evil)*

Adapted from Szekely, E.B. (1994). *Archeosophy, A New Science.*
U.S.: International Biogenic Society.

these four basic elements is then further divided into good and evil to produce a total of eight different fundamental elements (twelfth move of the Creator). Refer to Figure 24 for details.

Sumerian Cosmogony and the Constitution of the Universe

Based on these eight-fold elaborations of the four basic entities (Time, Force, Space and Matter), we can draw our second correspondence with the Constitution of the Universe. The first level of unfoldment in the Constitution of the Universe is the emergence of the eight prakritis or eight syllables of the first pada which arise from the first syllable of Rig Veda, "A-K" (refer to Figure 1). As we have already established, there appears to be a close resemblance between the first syllable of Rig Veda, "A-K," and the construction of the Cosmic Nebulae from the Creator or creative principle. We further propose that the eight fundamental prakritis of the Constitution of the Universe correspond to the eight-fold elaboration process on the Cosmic Nebulae. In each elaboration eight new lines are created giving rise to the eight seasons, eight types of energy, eight cardinal directions and the eight elements. This general eight-fold elaboration process of the four primal entities matches the eight prakritis of the Constitution of the Universe.

The next level of expansion in the Constitution of the Universe is the interpretation of each of the eight syllables of the first pada of Rig Veda with respect to rishi (knower), devata (process of knowing) and chhandas (known). This produces a total of twenty-four units. A similar view was also held by the ancient Sumerians. According to Szekely (Szekely, 1974; Szekely, 1976; Szekely, 1994) the ancient Sumerians, as well as the ancient Persians (refer to next section), considered that underlying everything was a basic three-fold natural code of life: "Good thoughts, good words and good deeds." This simplistic and holistic view explained the life of the individual and regulated the behavior of all types of relationships. Each of these three codes matches the interpretations of rishi, devata and chhandas, respectively. The first code, "Good thoughts," embodies qualities that relate most with the abstract nature of rishi: abstruse, subtle and non-expression. The second code, "Good words," exhibits characteristics that are more expressed than thoughts. Speaking and verbally communicating involve a transfer of information and feelings to the surroundings.

This matches the devata interpretation which is associated with motion, dynamism and expression. The last code, "Good deeds," is the most concrete and material expression of the three-fold code of life. This is equated with the chhandas interpretation which is attributed to the material or physical nature of an entity. These three codes of life, when viewed with respect to the eight-fold elaboration process of the four primal entities, comprise a total of twenty-four units, corresponding to the twenty-four units in the Constitution of the Universe.

The final stage of Sumerian Cosmogony involves the combination of the eight-fold elaborations on the four primal entities (Time, Force, Space and Matter) to create a universal tapestry of sixty-four squares. To accomplish this, four additional moves are required. The simplest way to comprehend this last stage is to refer to Figure 24 which geometrically depicts this process. To start, we will first combine the eight seasons of Time with the eight energies of Force. Geometrically this means that we are simply joining or connecting the lines for the seasons with the lines for energies. This process consists of two parts. The first part (representing the thirteenth move of the Creator) joins the four lines to the right of Time with the four lines to the right of Force. The second part (fourteenth move of the Creator) joins the four lines to the left of Time with the four lines to the left of Force. This integration of the eight aspects of Time with the eight aspects of Force represents the positive and negative energies during the eight seasons.

In the next step the eight cardinal points of Space are combined with the eight elements of Matter. This process is similar to the procedure just described for the integration of the eight seasons with the eight energies. In this case, however, we are connecting horizontal lines instead of vertical lines. The first step (fifteenth move of the Creator) connects the four lines above Space with the four lines above Matter. The second step (sixteenth and final move of the Creator) joins the four lines below Space with the four lines below Matter.

When viewed together, the integration of the elaborated aspects for Time, Force, Space and Matter constitutes a framework or grid of sixty-four squares (refer to Figure 24). This represents the geometrical foundation upon which the universe is structured. There is one more step required – dividing these sixty-four squares into good and evil. This is accomplished by alternating the color of the squares to produce what looks like the modern-day chessboard with thirty-two black squares (forces of darkness) and thirty-two white squares (forces of light). This concept of duality represents the basic paradigm of how the ancient Sumerians viewed the universe. They believed that every object we observe must have a shadow or a negative counterpart which is inseparable from the positive value. Thus there is a continuous conflict between the forces of good and the forces of evil. Figure 24 displays the final tapestry of creation according to the ancient Sumerians.

At this stage we can draw our complete correspondence between Sumerian cosmogony and the Constitution of the Universe. The second level in the Constitution of the Universe is the elaboration of the eight syllables of the first pada into sixty-four units. Likewise, according to Sumerian cosmogony, the eight-fold elaborations of the Cosmic Nebulae merge to create the final tapestry of creation consisting of a framework of sixty-four squares (refer to Figure 24). Further, each of these sixty-four aspects of the cosmic tapestry can be viewed with respect to the three-fold code of life, thereby creating a total of 192 different flavors. This matches the last level of the Constitution of the Universe template in which the sixty-four syllables of Rig Veda are interpreted with respect to rishi, devata and chhandas (a total of 192 units).

The Game of "Asha"

In concluding this section, we would like to describe a yearly ceremony called the "Day of Atonement" or "Game of Asha," which was performed in Sumerian times (Szekely, 1994). "Asha" is a word meaning "Cosmic Order." The "Game of Asha" was a drama or ritual that took place once a year on a large platform constructed upon a framework of sixty-four squares, around which the whole community gathered. The platform of sixty-four squares was a representation of the universe woven out of time, space, force and matter. This game or ritual was regarded as the "Game of Kings" and the "King of Games," and is linked to the game we know in modern times as chess.

Each game of Asha or chess was regarded as a reflection of the eternal cosmic drama between the forces of light and the forces of dark. This yearly drama or ritual was essentially a game of chess played by real individuals who represented cosmic forces. The purpose of the ceremony was to remind the citizens of the function of Cosmic Order in daily life and to inspire them in the quest for perfection and for life in accord with the forces of light. In the next section we expound upon the ancient Persian understanding of creation, which was based on the principles of the game of chess and the structure of the Constitution of the Universe.

Section II

Ancient Persia – Zend Avesta of Zarathustra

The Zend Avesta is the sacred writings of Zoroastrianism, the ancient Persian religion founded by the prophet Zarathustra (meaning "shining star" or "star-worshipper"). Unfortunately, about three quarters of the original Zend Avesta is believed to have been lost, or destroyed by Alexander the Great in his conquest of Persia. The surviving one third of this text was assembled and written down about one thousand years ago. Prior to that time the knowledge of the Zend Avesta was handed down in an oral tradition by Iranian priests. Apart from the work of Edmond Szekely (Szekely, 1957, 1974, 1976, 1990, 1994) there appears to be very little accomplished regarding the understanding of the Zend Avesta of Zarathustra. In addition to summarizing the elegant work of Szekely, we also demonstrate that the understanding of the origin of creation according to the Zend Avesta is the same as the numerical structure of the Constitution of the Universe.

The Zend Avesta is the Sacred Book of Life

Although there is some controversy over the exact date and location of the prophet Zarathustra, Greek records tell us that Zarathustra was older than Plato by 6,000 years and resided somewhere in ancient Persia (modern-day Iran) and/or Mesopotamia (Settegast, 1987). It is believed that Zarathustra received his wisdom from "Ahura Mazda," the supreme God, from Whom he produced his famous book, the Zend Avesta. Interestingly many of the worldviews and teachings in the Zend Avesta are also found in neighboring civilizations such as Vedic India, Sumeria and Greece. In fact, the Vedas of India and the Zend Avesta use languages that are similar to each other. For example, both reference a sacred plant which was believed to give spiritual power and exhilaration – in Iran it was called "haoma" and in Vedic India it was called "soma" (Szekely, 1990).

The Zend Avesta (translated as "Book of Life") is a sacred book drawing its knowledge from the wisdom of the natural environment. It is a universal encyclopedia containing chapters which deal with astronomy, psychology, philosophy, health, gardening and a whole range of subjects (Szekely, 1990). At the center of the Zend Avesta is the teaching that divine love was the greatest cosmic power and that it is natural for humans to express this love. Underlying everything, the ancient Persians had a three-fold natural code of life: "Good thoughts, good words and good deeds" (Szekely, 1974). As long as the individual followed this simple and natural code, there was no need for the myriad of civil laws that we commonly find in modern countries.

Living in balance with the forces of nature was the key to individual development and was considered to be a way of life that was achievable only by one's self. The individual who lived in harmony with the forces of nature was believed to enjoy an active creative life that would bring the highest measure of happiness and service to others. On the other hand, the individual who deviated from the forces of nature would find life becoming more miserable and frustrating (Szekely, 1974). Thus the purpose of the Zend Avesta was to show which forces of nature were friendly and which were unfriendly. In addition, it teaches how to cooperate with the friendly forces and how to avoid contact with the unfriendly ones.

Our Universe is Governed by the Positive and Negative Aspects of Sixteen Forces

According to the Zend Avesta, there are sixteen forces, eight cosmic and eight natural, representing the sum total of the universe. Zarathustra taught his disciples that the best thing in life was to be in harmony with these sixteen forces of nature arising from Asha, the Cosmic Order. Humans were considered to be the center of the universe with the forces of nature situated above and below. The mechanics by which these forces of nature are created is described in Fargard I, Vendidad section of the Zend Avesta (Szekely, 1994).

Similar to Sumerian cosmogony (refer to last section), Zoroastrianism believed that the origin of the universe started from a point. This point then becomes divided into thirty-two paragraphs or verses representing the positive and negative aspects of the sixteen forces of nature. These thirty-two verses constitute the Fargard I, Vendidad section of the Zend Avesta. Each force of nature has a name and a corresponding symbol (Szekely, 1994).

One of the clearest ways Zarathustra depicted the forces of nature governing the universe was to think of them as chess pieces operating in a world represented by the chess board of sixty-four squares. Further, the interaction of the various forces of nature were modeled by the movement and rules of the game of chess. In this way, creation was viewed as one macroscopic "game" governed by the forces of nature. As we will discuss later, the game of chess, as we know it today, may have originated from the Zend Avesta.

The sixteen positive forces of nature are located on one side of the chess board and on the other side are situated the corresponding sixteen negative forces of nature. The arrangement of these thirty-two forces is shown in Figure 25. In row "A" the eight positive cosmic forces, also known as the invisible or spiritual forces are located. It was believed that one must live in harmony with these eight cosmic forces prior to living in harmony with the eight natural forces. The "ways" of the cosmic forces were said to be mysterious and to reveal themselves only to those who devoted their lives to know them. On the opposite end of the board, row "H" in Figure 25, the eight negative cosmic forces are located. These eight negative cosmic forces are the counterparts or "shadows" of the eight positive cosmic forces.

In row "B" the eight natural or visible forces of nature are situated. As with the cosmic forces, each natural force has a negative counterpart at the opposite end of the board located in row "G." The eight natural forces represent entities residing in the physical or material world.

The sixteen positive forces create what is known as the Kingdom of Ahura Mazda (the kingdom of light). In this Kingdom the eight cosmic forces are often referred to as the eight "Ahuras" and the eight natural forces are commonly known as the eight "Fravashis." Opposing the Kingdom of Ahura Mazda are the sixteen negative forces which constitute the Kingdom of Angra Mainyu (the kingdom of darkness). In this Kingdom the eight negative cosmic forces are often referred to as the eight "Devas" and the eight negative forces as the eight "Khrafstras" (Szekely, 1974, 1994).

Description of the Eight Cosmic Forces

All the forces of nature operate under well-defined conditions and restrictions. These laws constitute the rules of the game of creation, or the game of chess.

Figure 25. Zarathustra's View of Creation as a Macroscopic "Game" of Chess

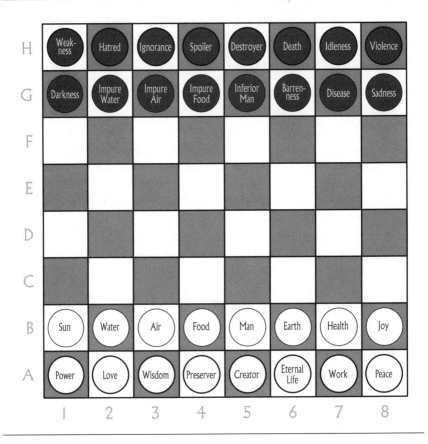

Reproduced with modifications from Szekely (1974 and 1987).

In the following list we describe each of the eight cosmic forces, the movements they are allowed to make, and their opposing negative cosmic forces.

- The first cosmic force to manifest, called the "Creator," is symbolized by a small white circle. To be in harmony with the Creator, we should always do things that are creative and original. Counteracting the Creator is the negative cosmic force called the "Destroyer," represented by a small black circle. On the chess board both these forces can move only one square at a time in any direction.

- The second cosmic force, the "Preserver," is symbolized by a circle divided into two equal parts by a curved line (this is also the symbol of the Yin/Yang of ancient China). To be in balance with the Preserver means that we should prevent any good thing or value from deterioration. (In our present age of colossal waste we are far from this ideal.) Opposing the Preserver is the negative cosmic force called the "Spoiler," which has the same symbol as the Preserver except that it is black. Both these forces have great versatility on the board and can move any distance and in any direction. These entities are considered to be the most powerful forces.

- The third cosmic force to emerge, "Eternal Life," is symbolized by a circle divided into eight curved spokes. To live in accord with this force, we are required to be sincere toward ourself and others. The counterbalance of this force, called "Death," has the same symbol as Eternal Life except that it is black. Both these forces move in diagonal lines on the chess board (similar to bishop's move in modern chess). Movement in an oblique direction represents "bent on victory."

- The fourth cosmic force, "Wisdom," is symbolized by a white circle about the size of a point. This force is considered to be the "light of life," and to be in harmony with it we must understand the cosmic order and our role in it. Opposing this force is the negative cosmic force called "Ignorance," symbolized by a black dot. Both forces move diagonally similar to the previous pair of cosmic forces.

- The fifth cosmic force, "Work," is symbolized by an eight-pointed star representing the worker of the sky and the universe. To be in balance with this force, one must perform his/her daily tasks with care, efficiency and a deep feeling of satisfaction. Put another way, we should take pride in our work and, if work is worth doing, it is worth doing well. This is not the type of work we may be familiar with; it is creative work (the science of inventions) which allows us to continue the work of God on this planet. The opposite force to Work, called "Idleness," is the same symbol as for Work except it is black. Both forces move in the same way that the Knight piece moves in modern chess.

- The sixth cosmic force, "Love," is symbolized by a white triangle. Living in accord with this force is connected to the previous force of work because love is manifested in work. Living love means kind and gentle words to others. Opposite the force of Love is "Hatred," symbolized by a black triangle. Both forces move in the same way as the previous pair of forces. It is interesting that several proverbs exist which encapsulate the essence of the forces Work and Love. For example, there is an ancient saying in the Bible "Love is stronger than Death," and a Roman proverb "Work overcomes all evil."

- The seventh cosmic force, "Peace," is symbolized by a white crescent moon. Living in accord with peace means to establish inner peace and to express this inner peace to help improve the outer environment. Counterbalancing this force is "Violence," symbolized by a black crescent moon. Both forces move any distance in the vertical or horizontal directions, which is symbolic of the sweeping motions of infinite strength.

- The last cosmic force, "Power," is symbolized by a white square. This force represents the iron law of cause and effect and states that everything which happens to us is the result of our past deeds. Opposing this force is the negative cosmic force called "Weakness," symbolized by a black square. Both forces move in the same way as the previous pair of forces.

In Table 10 we present Szekely's translation of verses 1 through 16 of Fargard I, Vendidad of the Zend Avesta, which describe the creation of the foregoing sixteen positive and negative cosmic forces. (Table 11 presents verses 17 through 32.)

Description of the Eight Natural Forces

All the natural forces move under the same rules and restrictions. Each natural force has the same movements as a "pawn" in modern chess. If a natural force reaches the opposite side of the board, it can choose to become any cosmic force it desires, with the exception of the Creator and Destroyer. The following describes each of the eight positive natural forces and their corresponding opposite negative forces (Szekely, 1974, 1994):

- The first natural force, called the "Sun," is symbolized by a central point with eight rays emanating from it. This force represents the science of heliotherapy and is the source of energy and power for

Table 10. E.B. Szekely's Translation of Verses 1-16 of Fargard I, Vendidad of the Zend Avesta: Creation of the Eight Ahuras and the Eight Devas[a]

Verse 1
"Ahura Mazda
Spake unto Spitama Zarathustra saying:
First I have made the Kingdom of Light,
Dear to all Life."
Ahura Created: Creator[b]; Position: A5[c]

Verse 2
"I cast my shadow, Angra Mainyu,
Who am all Death."
Deva Created: Destroyer; Position: H5

Verse 3
"The second of the good Kingdoms
Which I, Ahura Mazda, created,
Was the Ahura of Preserver."
Ahura Created: Preserver; Position: A4

Verse 4
"Thereupon came Angra Mainyu,
Who is all Death,
And he counter-created the Deva of Spoiler."
Deva Created: Spoiler; Position: H4

Verse 5
"The third of the good Kingdoms
Which I, Ahura Mazda, created,
Was the Ahura of Eternal Life."
Ahura Created: Eternal Life; Position: A6

Verse 6
"Thereupon came Angra Mainyu,
Who is all Death,
And he counter-created the Deva of Death."
Deva Created: Death; Position: H6

Verse 7
"The fourth of the good Kingdoms
Which I, Ahura Mazda, created,
Was the Ahura of Wisdom."
Ahura Created: Wisdom; Position: A3

Verse 8
"Thereupon came Angra Mainyu,
Who is all Death,
And he counter-created the Deva of Ignorance."
Deva Created: Ignorance; Position: H3

Verse 9
"The fifth of the good Kingdoms
Which I, Ahura Mazda, created,
Was the Ahura of Work."
Ahura Created: Work; Position: A7

Verse 10
"Thereupon came Angra Mainyu,
Who is all Death,
And he counter-created the Deva of Idleness."
Deva Created: Idleness; Position: H7

Verse 11
"The sixth of the good Kingdoms
Which I, Ahura Mazda, created,
Was the Ahura of Love."
Ahura Created: Love; Position A2

Verse 12
"Thereupon came Angra Mainyu,
Who is all Death,
And he counter-created the Deva of Hatred."
Deva Created: Hatred; Position: H2

Verse 13
"The seventh of the good Kingdoms
Which I, Ahura Mazda, created,
Was the Ahura of Peace."
Ahura Created: Peace; Position A8

Verse 14
"Thereupon came Angra Mainyu,
Who is all Death,
And he counter-created the Deva of Violence."
Deva Created: Violence; Position: H8

Verse 15
"The eighth of the good Kingdoms
Which I, Ahura Mazda, created,
Was the Ahura of Power."
Ahura Created: Power; Position A1

Verse 16
"Thereupon came Angra Mainyu,
Who is all Death,
And he counter-created the Deva of Weakness."
Deva Created: Weakness; Position: H1

[a] Szekely, E.B. (1994). *Archeosophy, A New Science.* USA: International Biogenic Society, pp. 30-35.
[b] Ahura/Deva Created = What force of nature is being described by the verse.
[c] Position = The position of that force on the cosmic framework of 64 squares *(refer to Figure 25).*

Table 11. E.B. Szekely's Translation of Verses 17-32 of Fargard I, Vendidad of the Zend Avesta: Creation of the Eight Fravashis and the Eight Khrafstras[a]

Verse 17
"The ninth of the good Kingdoms
Which I, Ahura Mazda, created,
Was the Fravashi of the Sun."
Fravashi Created: Sun[b]; Position: B1[c]

Verse 18
"Thereupon came Angra Mainyu,
Who is all Dealth,
And he counter-created the Khrafstra of darkness."
Khrafstra Created: Darkness; Position: G1

Verse 19
"The tenth of the good kingdoms
Which I, Ahura Mazda, created,
Was Fravashi of Water."
Fravashi Created: Water; Position: B2

Verse 20
"Thereupon came Angra Mainyu,
Who is all Death,
And he counter-created the Khrafstra
Of Impure Water."
Khrafstra Created: Impure Water; Position: G2

Verse 21
"The eleventh of the good Kingdoms
Which I, Ahura Mazda, created,
Was the Fravashi of Air."
Fravashi Created: Air; Position: B3

Verse 22
"Thereupon came Angra Mainyu,
Who is all Death,
And he counter-created the Khrafstra
Of Impure Air."
Khrafstra Created: Impure Air; Position: G3

Verse 23
"The twelfth of the good Kingdoms
Which I, Ahura Mazda, created,
Was the Fravashi of Earth."
Fravashi Created: Earth; Position: B6

Verse 24
"Thereupon came Angra Mainyu,
Who is all Death,
And he counter-created the Khrafstra
Of Barrenness."
Khrafstra Created: Barrenness; Position: G6

Verse 25
"The thirteenth of the good Kingdoms
Which I, Ahura Mazda, created,
Was the Fravashi of Food."
Fravashi Created: Food; Position: B4

Verse 26
"Thereupon came Angra Mainyu,
Who is all Death,
And he counter-created the Khrafstra
Of Impure Food."
Khrafstra Created: Impure Food; Position: G4

Verse 27
"The fourteenth of the good Kingdoms
Which I, Ahura Mazda, created,
Was the Fravashi of Health."
Fravashi Created: Health; Position B7

Verse 28
"Thereupon came Angra Mainyu,
Who is all Death,
And he counter-created the Khrafstra of Disease."
Khrafstra Created: Disease; Position: G7

Verse 29
"The fifteeneth of the good Kingdoms
Which I, Ahura Mazda, created,
Was the Fravashi of Man."
Fravashi Created: Man; Position B5

Verse 30
"Thereupon came Angra Mainyu,
Who is all Death,
And he counter-created the Khrafstra
Of Inferior Man."
Khrafstra Created: Inferior Man; Position: G5

Verse 31
"The sixteenth of the good Kingdoms
Which I, Ahura Mazda, created,
Was the Fravashi of Joy."
Fravashi Created: Joy; Position B8

Verse 32
"Thereupon came Angra Mainyu,
Who is all Death,
And he counter-created the Khrafstra of Sadness."
Khrafstra Created: Sadness; Position: G8

---

[a] Szekely, E.B. (1994). *Archeosophy, A New Science.* USA: International Biogenic Society, pp. 30-35.
[b] Fravashi/Khrafstra Created = What force of nature is being described by the verse.
[c] Position = The position of that force on the cosmic framework of 64 squares *(refer to Figure 25).*

all life. Opposing this force is "Darkness," which has the same symbol as the Sun except that it is black.

- The second natural force, called "Water," is symbolized by a white fish. Water is an essential element of life and is represented in living organisms as blood. Opposing this force is "Impure Water," symbolized by a black fish.

- The third natural force, called "Air," is symbolized by a white bird. In the Zend Avesta breathing is considered to be the greatest source of energy and plays a tremendous role in health. The texts further prescribe that one should spend lots of time outdoors – breathing and life go together. In the words of Szekely, "Where is Breath, there is Life. Where is Life, there is Breath. Breathing is Life" (Szekely, 1974, p. 19). The counterbalancing force to Air is "Impure Air," symbolized by a black bird.

- The fourth natural force, called "Earth," is symbolized by a white earthworm. This force describes the science of horticulture, fertility, sexual energies and the creative power of humankind. Opposing this force is "Barrenness," symbolized by a black earthworm.

- The fifth natural force, called "Food," is symbolized by a white ear of wheat. Wheat formed a central part of the diet for people in the Middle East and represented a vital force of nutrients. Opposing this force is "Impure Food," symbolized by a black ear of wheat.

- The sixth natural force, called "Health," is symbolized by a white tree. The tree was a symbol for the medical sciences and the wood of the tree was used to create an altar of fire which represented health and vitality. The counterbalancing force to Health is called "Disease," symbolized by a black tree.

- The seventh natural force, called "Man," is symbolized by a combination of three symbols: a triangle, a square and a circle. The triangle represents the legs and corresponds to the cosmic force of Love; the square delineates the body and matches the cosmic force of Power; and the circle describes the head, which corresponds to the cosmic force of Wisdom. Man is the representation of the Creator on this planet and he is considered to be the measure of all things. Placed directly in front of the cosmic force of the Creator, Man occupies the most glorious position on the board. In addition, Man was considered to be the master of life and not just a part of nature like animals. The opposite force of Man is the negative natural force called the "Inferior Man," which has the same symbol as Man except that it is black.

- The last natural force, called "Joy," is symbolized as a white flower. To live in accord with this natural force, we should always feel and express joy in everything we do. The opposite force, called "Sadness," is symbolized by a black flower.

In Table 11 we present Szekely's translation of verses 17 through 32 of Fargard I, Vendidad of the Zend Avesta which describe the creation of the foregoing sixteen positive and negative natural forces.

Representing the forces of nature and their interactions in the form of a chess game brings the ideas of universal functioning into daily living. The game symbolizes and exemplifies the positive and negative forces which are eternally balancing and counteracting each other. The individual should aspire to live in harmony with the sixteen positive forces and should have a clear understanding of his or her place and role in the universe.

The Legend of the Game of Chess

How the game of chess became used by Zarathustra as a tool to display the mechanics of the universe is the subject of an amusing legend (Szekely, 1976). It is said that long ago King Vistaspa of Persia became bored with life and asked his people to find something that would interest him. No one was successful until Zarathustra presented him with the game of chess. In teaching King Vistaspa the game of chess, Zarathustra demon-strated to the king all the laws of the universe and of life. The object of the game was to capture the Creator or Destroyer. This renewed the king's interest in life. As a reward the king granted Zarathustra anything he desired. Zarathustra used this opportunity to teach the king. He asked the king to give him one grain of wheat for the first square on the chess board, two grains for the second square, four grains for the third square, eight grains for the fourth square, and continue until all sixty-four squares were used up. The king laughed at this apparently foolish and modest request and told his men to calculate the quantity and deliver it to Zarathustra. Much to his surprise, however, he soon realized that the amount he had promised Zarathustra was equal to 18,446,744,673,709,551,615 grains of wheat! This is enough wheat to cover the surface of the Earth nine inches deep. Consequently the king apologized to Zarathustra for his inability to fulfill the promise.

The Zend Avesta is Identical to the Structure of the Constitution of the Universe

So far we have given a brief overview of the teachings of Zarathustra as recorded in the Zend Avesta. Using this wisdom, we will now proceed to locate Zarathustra's descriptions of the mechanics of creation in the numerical structure of the Constitution of the Universe. The first unit in the Constitution of the Universe is represented in the Rig Veda by the syllable "A-K." The letter "A" contains in seed form the complete knowledge of natural law which collapses to the point value of "K." This collapse occurs in eight steps that are subsequently responsible for the first eight syllables of the first pada of Rig Veda (refer to Figure 1). We propose that these eight steps match the creation of the eight cosmic or invisible forces of nature in the Zend Avesta.

In the initial stage of expression of the Constitution of the Universe the first syllable "A-K" unfolds into the eight syllables or prakritis of the first pada (refer to Figure 1). Similarly in the Zend Avesta, the eight natural or visible forces arise after the creation of the eight cosmic forces of nature. These eight natural forces correspond to the eight prakritis. This connection appears particularly apt considering that four of the natural forces are called "Air," "Sun," "Water" and "Earth" which exactly correspond with the following prakritis: Vayu (translated as air); Agni (translated as fire); Jala (translated as water); and Prithivi (translated as earth). Although the match between the remaining four natural forces and the other prakritis is less clear, we feel these parallels provide qualitative support to the proposed correspondence between the eight natural forces and the eight prakritis.

From the eight prakritis of Rig Veda, the next stage of expansion in the Constitution of the Universe is the interpretation of each prakriti with respect to rishi (knower), devata (process of knowing) and chhandas (known). This produces a total of twenty-four units. A similar view is also present in the Zend Avesta. According to Szekely, Zarathustra, as well as the ancient Sumerians (refer to previous section), considered that underlying everything was a basic three-fold code of life: "Good thoughts, good words and good deeds." As described earlier, this simple and holistic view explained the life of the individual and regulated the behavior of all types of relationships. The first code, "Good thoughts," embodies characteristics such as subtle, abstruse and non-expressed which correspond to the abstract and refined nature of rishi. The second code, "Good words," is more expressed than thoughts. Speaking and verbally communicating involve the transfer of information and feelings to the environment. This matches the devata interpretation which is associated with motion, dynamism and expression. The last code, "Good deeds," is the most concrete and material expression of the three-fold code of life. This is equated with the chhandas interpretation which is attributed to the material or physical nature of an entity. Applying or interpreting each of these three codes with respect to each of the eight natural forces produces a total of twenty-four units corresponding to the twenty-four units of the Constitution of the Universe.

In the second stage of elaboration of the Constitution of the Universe the eight prakritis of Rig Veda expand into sixty-four units. These sixty-four units exactly match the sixty-four squares of the chess board. Through this tapestry of sixty-four squares the interaction of the cosmic and natural forces is able to display the mechanics of creation. The final correspondence with the Constitution of the Universe stems from the sixty-four basic syllables of Rig Veda interpreted with respect to rishi, devata and chhandas, thus comprising a total of 192 units. Likewise each of the sixty-four squares of chess, that accommodate the interaction of the thirty-two forces of nature, can be described with respect to the three-fold code of life (Good thoughts, good words and good deeds), thereby making a total of 192 units.

Chess is the "Game of Kings" and the "King of Games"

In conclusion, we would like to provide a brief glimpse of the importance of the game of chess. Chess is one of the oldest games known to humankind. Although the game may have arisen in ancient Persia, it appears that numerous other civilizations may have also independently developed the game. First, oriental cultures have a long history of the chess, and remnants have been found throughout the Chinese, Japanese, Burmese and Siamese cultures. Second, the game of chess is found on the walls of Egyptian temples and tombs and was associated with Iris, the wife of Osiris. Third, description of the game of chess is often found in the Vedic literature where it is associated with various gods. Fourth, chess was actively played in Russia and by notable leaders such as Lenin and Karl Marx (Falkner, 1994; Murray, 1962).

In most of the ancient civilizations, however, chess was not a game for the common citizen. It was one of the guarded secrets of society and was reserved for priests, kings and philosophers. The game was considered to have enormous educational and spiritual importance and was used in the following realms of life: teaching, administration, decision-making, mental training, military strategy, complex mathematics, divination, astronomy and astrology (Falkener, 1961; Murray, 1962; Patel, 1992). For these reasons chess is often referred to as the "king of games" and the "game of kings."

As with other games, chess was not created for mere play or to pass time. It provided a way for mental relaxation and culturing of the mind (Pennick, 1989). In fact, some believe that the invention of chess occurred during the time of war. Often mastery of chess is held to be higher than the mastery of other board games. Interestingly, chess has been used in some countries to improve memory. For example, in Caracas, the capital city of Venezuela, chess was approved to be a part of the school curriculum, and after several months of its implementation, tests demonstrated that children had increased IQ scores.

We conclude this section with two quotes regarding the game of chess:

> *"Man is not alone on the chessboard of Life.*
> *He is surrounded by Divine Powers, Love and*
> *Wisdom and all the good forces of Providence*
> *in this world of Shadows and Lights."*
>
> – Tolstoy
> (Szekely, 1974)

> *"And we are Pieces of the Game He plays*
> *Upon this Chequer-board of Nights and Days;*
> *Hither and thither moves, and checks, and wins,*
> *And then another Cosmic Game begins."*
>
> – Omar Khayyam,
> 11th century Persian Poet
> (Szekely, 1976, p. 31)

SECTION III

ANCIENT ISRAEL – THE KABBALAH

The Kabbalah (a Hebrew word meaning "to receive" or "tradition") is the inner or mystical teachings of Judaism. Although its origins are ancient, arising out of Israel, it has been preserved by esoteric Judaism and remains a living tradition belonging to the mainstream of traditional Jewish practice. At the center of the Kabbalah is what is known as the Tree of Life, describing the perennial teachings about the attributes of God, the universe and man. The Tree of Life is a diagram which displays the objective laws that govern the universe and is an image of Divine Unity. Further, the Kabbalah is the underlying basis of the Judaic-Greco-Christian tradition. In this section we will discuss the foundational principles of the Tree of Life in the Kabbalah and demonstrate an interesting parallel with the numerical structure of the Constitution of the Universe.

Historical Perspective

Although the origins of the Kabbalah are historically unknown, it is traditionally believed to go back to the time of Abraham, but it may even predate him by several millennia. It is thought that the wisdom of the Kabbalah was imparted by divine revelation and has been handed down over the centuries by a discrete tradition. During this time the Kabbalah has accumulated a remarkably rich and wide array of interpretations. Although the Holy Bible does not explicitly discuss the Kabbalah, the symbol for the Kabbalistic Tree of Life is found in a number of different places in the Bible. For example, the Tree of Life symbol underlies the seven-branched candlestick (Menorah) given to Moses by God on Mount Sinai (Exodus 25) and the ten commandments are said to have sprung from the wisdom inherent in the Kabbalah.

As each new generation has emerged, the Kabbalah has been constantly reformulated to remain understandable. To this day Kabbalists continue to live by its doctrine and principles of life. The Kabbalah views humans and all organisms as microcosms reflecting the same order and patterns of the macrocosm of the universe. This view is expressed in the image of the Kabbalistic Tree of Life which is a model of creation.

Our Creation Begins with God

According to the Kabbalah, creation begins with God. God is a state of Unity and is described as the One unmanifest reality encompassing everything. In the words of Halevi (Halevi, 1979, p. 5), "God is God. There is no thing to compare with God. God is God." Out of this One reality Kabbalists have recognized that God has three aspects: "Ayin," "Ayin Sof" and "Ayin Sof Or" (Halevi, 1979).

In Hebrew the first aspect of God, Ayin, is translated as meaning "No Thing" or "Nothing." This Transcendental aspect of God describes God as being absolute nothing and as residing beyond existence. From Ayin arises the second aspect of God, Ayin Sof, which is translated as meaning "Without End." Ayin Sof depicts the characteristic of God that is found everywhere, the totality of what is and what is not. It is the value which transcends all that exists. Another way to understand Ayin Sof is to think of it as analogous to the sap of a tree (Ponce, 1973). Just as the sap of a tree pervades all aspects of the branches, leaves and roots, so too does the Ayin Sof aspects of God pervade all that there is in the universe. The third aspect of God is called Ayin Sof Or which means "Endless Light." It is from this aspect of God that the manifest universe originated.

A Beam of Divine Will
Originates from Zimzum

According to the oral tradition of the Kabbalah, the pivotal reason for the origin of existence is that "God wished to behold God" (Halevi, 1979). Put another way, creation began only because of the free will and desire of God. However, for this to occur, God had to withdraw or contract the Absolute All, or Ayin Sof, so that a void could appear. By creating a void, a space for the genesis of the creation was constructed. Without such a contraction there would have been no room for the universe to come into being (Sheinkin, 1986). This process of withdrawal and contraction is known as Zimzum. It is important, however, to realize that this idea of "withdrawal" or "contraction" of the omnipresent value of God is simply a covering up, or hiding of, Ayin Sof so that multiplicity can emerge. In fact, Kabbalists believe that there is never any real withdrawal, but that it is just an intellectual conception. For further discussion of this concept of the hidden God, the reader is referred to the work of Scholem (Scholem, 1965 and 1978).

After this act of withdrawal, in which all energies have become concentrated into a central arena, an expansion must occur. In this expansion a beam or ray of energy and light emerges from the Endless Light of Ayin Sof Or, surrounding the void (Ponce, 1973). This beam of Divine Will manifests in ten distinct stages of Emanation or ten Divine Utterances, which have been known since the Middle Ages as the "Sefiroth" (Sefira, singular) or the Kabbalistic Tree of Life. The Sefiroth, organized into a specific archetypal pattern, is the template for everything that was, is and will be.

This initial process of withdrawal of the Absolute All of God, known as Zimzum, to a beam of Divine Will matches with the genesis of the first syllable of the Constitution of the Universe, "A-K." The first letter, "A," represents a state of complete unity or unbounded intelligence corresponding to the concept of Ayin Sof which describes God as being omnipresent. The second letter, "K," representing a point value expression of "A," matches the emergence of the beam of Divine Will. Together the first two letters (first syllable) of Rig Veda, "A-K," describe the collapse of all possibilities to a point paralleling the act of Zimzum in which the Absolute All aspect of God is contracted to a beam of energy and light.

Description of the Ten Sefiroth

Most, if not all, Kabbalistic doctrine is based on the ten Sefiroth or Tree of Life, which resembles a perfect configuration of Divine Attributes. Everything that we can imagine or contemplate is present in the interactions and primal movements of the Tree of Life. Each Sefira contains all the others and together they are responsible for the laws governing our universe. Often they are viewed as basic forces or manifestations of particular deities that act as the tools for maintaining creation.

There are various methods for arranging or distributing the ten Sefiroth on the Tree of Life. Of these different arrangements there appears to be no one way that alone is considered correct or incorrect, because each classification serves a different purpose for Kabbalists. In Figure 26 we have presented the most common arrangement, known as the Lurianic scheme. This system is based on the work of Isaac Luria who lived in Palestine during the sixteenth-century and studied at the Kabbalistic center of Safed (Halevi, 1979). The transliteration of the Hebrew words for each Sefira has also been the subject of much debate. Moreover, different scholars have their own systems for spelling each of the ten Sefiroth. In this section

Figure 26. Manifestation of Creation According to the Kabbalah: The Three Aspects of God and the Common Order of the Ten Sefiroth (Tree of Life)

Three Aspects of God:

- AYIN – Absolute Nothing, Transcendental Aspect
- AYIN SOF – Absolute All

  ↓ **Zimzum** – Contraction or Withdrawal

- AYIN SOF OR – Endless Light

  ↓ Divine Will

Column of Equilibrium or Consciousness

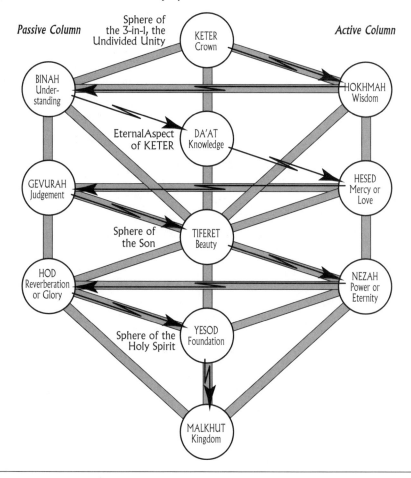

References: Lightning Flash – Halevi (1979) pp. 6-7; Ten Sefirot, KETER to MALKHUT – Scholem (1978) p. 106; AYIN, AYIN SOF and AYIN SOF OR – Halevi (1979) pp. 5-6; Holy Trinity of Unity—Son—Holy Ghost – Fortune (1984) p. 51.

we use the translations based on the work of two authors, G.G. Scholem and Z.S. Halevi, who are generally accepted to have written authoritative accounts on the subject of the Kabbalah.

As described earlier, the beam of Divine Will emanating from the Ayin Sof Or aspect of God manifests in terms of the ten Sefiroth. These ten Sefiroth can be visualized as a "lightning flash" (Halevi, 1979) or a zigzagging line (refer to the dark arrows in Figure 26). As a result of this process, the ten Sefiroth become distributed into three primary columns or pillars (see Figure 26). The central column represents consciousness or equilibrium. On either side of this central pillar are the other two columns representing opposite values. The left column is described as passive, severity, negative, feminine and mother. Opposing these qualities are the matching characteristics of the right column: active, mercy, positive, masculine and father, respectively. It is interesting to note that these three columns have a close resemblance to similar aspects in the I Ching of ancient China. The two opposite columns match the concept of yin and yang, while the middle pillar of consciousness can be equated to the Tao which lies between the yin and yang. (Refer to Chapter 8, Section IV for more discussion of the I Ching).

Connecting the ten Sefiroth are twenty-two paths or ways of life (refer to Figure 26). One of the significant aspects of these twenty-two paths is that they can be associated with the twenty-two letters of the Hebrew alphabet (Feuerstein, 1994). Often Kabbalists include the ten Sefiroth with the twenty-two paths, for a total of thirty-two. These thirty-two are considered to constitute the matrix of the universe.

Each of the ten Sefiroth and their commonly translated names are displayed in Figure 26, starting with Keter (meaning Crown) and ending with Malkhut (translated as Kingdom). Between the third Sefira (Binah) and the fourth (Hesed) there is something called "Da'at." Technically Da'at is not considered to be one of the ten Sefiroth, but is viewed as the external aspect of Keter, the first Sefira. Da'at is said to emerge out of nowhere and come directly from God (Scholem, 1978).

The Law of Octaves and the Constitution of the Universe

Since the relationships displayed in the Tree of Life underlie all existence, the properties of the Sefiroth can be understood and applied to any branch or discipline of knowledge. As a result, over the centuries Kabbalists have located aspects from numerous fields of knowledge in the structure of the Tree of Life. To locate the basic aspects of a discipline in the Tree of Life, Kabbalists often use what is known as the Law of Octaves. This fundamental law is formulated on the idea that a great string or monochord consisting of eight units stretches between the first and tenth Sefiroth (Keter and Malkhut; Halevi, 1972). These eight units correspond to the eight outer Sefiroth of the Tree of Life which are known as the Master Octave.

In Table 12 we have summarized the eight Sefiroth of the Master Octave and have provided two examples of how these eight Sefiroth can be associated with other disciplines. The first example is from music (Halevi, 1974, 1972). Each of the eight Sefiroth of the Master Octave is assigned one of the notes of the musical scale (from Do to Do). The second example is the assignment of the Sefiroth to the Planets (Halevi, 1974). In addition to these two examples Halevi (Halevi, 1972, p. 84) stated that, "The Law of Octaves besides being seen in the one great progression from Keter to Malkhut can be observed on every scale, including quite mundane phenomena such as the light spectrum, the periodic table of elements, and of course as music."

At this stage we are able to draw our second major correspondence between the Constitution of the Universe and the Kabbalistic Tree of Life. As described earlier, the first syllable of the Constitution of the Universe, "A-K," precisely matches the "withdrawal" of the Ayin Sof aspect of God to create the beam of Divine Will which manifests the ten Sefiroth. In the next level of the Constitution of the Universe the syllable "A-K" is elaborated into the eight syllables or prakritis of the first pada (refer to Figure 1). Likewise, the beam of Divine Will is expressed into the Master Octave that is embedded within the ten Sefiroth. We propose that the eight Sefiroth constituting the Master Octave match the eight prakritis of the Constitution of the Universe.

Further, it is worth mentioning two interesting points lending support to our proposed correlation between the Master Octave of the Tree of Life and the eight prakritis of the Constitution of the Universe. First, the eight musical tones Kabbalists have associated with the eight Sefiroth of the Master Octave (refer to Table 12) are the same eight tones which we matched with the eight prakritis of the Constitution of the Universe in Chapter 5. Second, an interesting parallel can be described from the association between the Sefiroth and planets of the solar system. In Table 12 we have summarized the specific Sefiroth in the Master Octave that correspond to specific planets (Halevi, 1974). As we examine the eight prakritis of Rig Veda (refer to Table 12), we notice that they can be divided into two groups: the first three representing subjective values and the last five representing the five elements. According to another aspect of the Vedic Literature called Jyotish (the science of prediction and cosmology), each of the five elements is associated with a particular planet (Parasara, 1984). The order of the planets corresponding to the last five prakritis match the order of the planets on our proposed correlation with the Sefiroth of the

Table 12. Proposed Correspondence Between the Eight Prakritis of the Constitution of the Universe and the Law of Octaves in the Sefiroth (Tree of Life)

Rig Veda:

AK	ni	mi	le	pu	ro	hi	tam

Prakriti:

Ahamkara	Buddhi	Manas	Akasha	Vayu	Agni	Jala	Prithivi

Translation:

Ego	Intellect	Mind	Space	Air	Fire	Water	Earth

Planets:

—	—	—	Jupiter	Saturn	Mars	Venus	Mercury

Sefiroth:

Keter	Hokhman	Binah	Hesed	Gevurah	Nezah	Hod	Malkhut

Meaning:

Crown	Wisdom	Understanding	Mercy	Judgement	Eternity	Reverberation	Kingdom

Musical Notes:

Do	Re	Mi	Fa	So	La	Ti	Do

Planets:

1st Swirlings	Zodiac	Saturn	Jupiter	Mars	Venus	Mercury	Earth

References:
1. Association between the last five Prakritis (five elements) and the Planets *(Parasara, 1984, Ch. 3)*.
2. The Sefiroth and their meanings *(Halevi, 1974, p. 20)*.
   Association between the Sefiroth and the musical notes *(Halevi, 1974, p. 25)*.
   Association between the Sefiroth and the planets *(Halevi, 1974, p. 134)*.

Master Octave. To develop a perfect match, all we need to do is to alter slightly the order of planets associated with the Sefiroth (i.e., to place Saturn in the place of Mars and shift Mars, Venus and Mercury over by one unit). Refer to Table 12 for clarification. Considering that the Kabbalah has survived the lapse of time and the flexibility in its interpretation, we feel that this last parallel provides additional support to an already strong correspondence between the eight Sefiroth of the Master Octave and the eight prakritis of the Constitution of the Universe.

Correspondence Between the Law of Triads and the Constitution of the Universe

The next level in the Constitution of the Universe is the interpretation of the eight prakritis with respect to rishi (knower), devata (process of knowing) and chhandas (known). This produces a total of twenty-four units. Likewise there is a similar concept in the Kabbalistic Tree of Life known as the Law of Triads. According to Halevi (Halevi, 1972), through the interaction of the Law of Triads and the Law of Octaves the relative universe came into existence.

The concept of a triad or trinity is prevalent throughout Judaism as it is in all religions. For example, the famous Jewish scholar Maimonides states that:

> *"The basic principle of all basic principles and the pillar of all sciences is to realize that there is a First Being who brought every existing thing into being. All existing things, whether celestial, terrestrial, or belonging to an intermediate class, exist only through His true existence."* (p. 43)

> *"His [First Being] Knowledge and His life are One, in all aspects, from every point of view, and however we conceive Unity … He is One, in every respect, from every angle, and in all ways in which Unity is conceived. Hence the conclusion that God is the One who knows, is known and is the knowledge (of Himself) – all these being One."* (p. 46)

> – Maimonides,
> as translated by Twersky, 1972

The last sentence of the quote is identical to the description of rishi (knower), devata (process of knowing) and chhandas (known). In the Tree of Life this notion of trinity or triads is amply represented. Each of the Sefiroth can form themselves into triads, and each of these triads is connected with the central pillar of consciousness. For example, the following three triads are widely known and discussed by Kabbalists:

- The triad consisting of Keter, Hokhmah and Binah;
- The triad consisting of Hesed, Gevurah and Tiferet; and
- The triad consisting of Nezah, Hod and Yesod.

These triads can be located by referring to Figure 26. However, one of the most important triads that appears to correspond to the Christian Trinity (i.e., the Father, the Son and the Holy Ghost) is located on the central column of consciousness itself. This triad consists of Keter, Tiferet and Yesod. Keter (or Da'at, the external aspect of Keter) is assigned the Sphere of the three-in-one which is the undivided Unity. Tiferet, lying in the center of Tree of Life image, represents the Sphere of the Redeemer or Son. Last, Yesod is assigned the Sphere of the Holy Spirit or the Enlightener (Fortune, 1984). We propose that this last triad naturally corresponds to the qualities of rishi, devata and chhandas. The rishi quality is equated with abstract intelligence, and we propose this matches Da'at, which expresses the undivided unity characteristics of Keter. The second interpretation is devata whose function is to be the connection or transition between rishi and chhandas. This corresponds to the Tiferet or the Sphere of the Son which connects the Holy Spirit with Father aspects of God. Last, the chhandas aspect is associated with the material or expressed nature of an entity. This matches Yesod or the Sphere of the Holy Spirit, the expressed spirit of God. When viewed with respect to each of these three Sefiroth, the remaining eight Sefiroth of the Master Octave produce a total of twenty-four units; these correspond to the twenty-four units of the Constitution of the Universe. Further support of this idea of viewing the eight (Sefiroth of the Master Octave) with respect to the three (Da'at, Tiferet and Yesod) is provided visually by looking at Figure 26. Each of the three Sefiroth making up the triad have paths radiating outwards connecting them to the eight Sefiroth located on the outside of the diagram. A summary of all the connections made so far between the Kabbalistic Tree of Life and the Constitution of the Universe is presented in Figure 27.

Figure 27. Locating the Numbers Eight and Three of the Constitution of the Universe in the Sefiroth (Tree of Life)

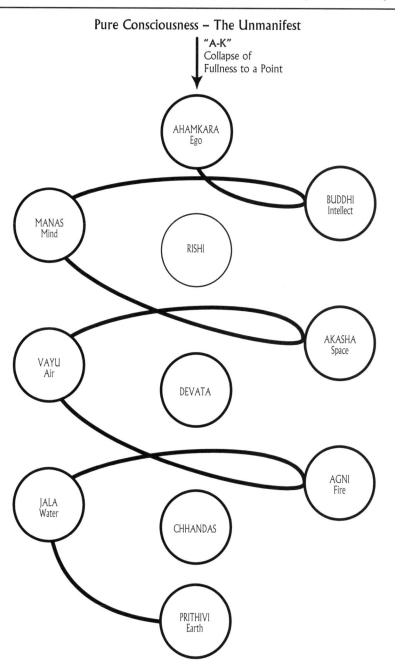

Pure Consciousness – The Unmanifest

"A-K"
Collapse of
Fullness to a Point

AHAMKARA
Ego

BUDDHI
Intellect

MANAS
Mind

RISHI

AKASHA
Space

VAYU
Air

DEVATA

AGNI
Fire

JALA
Water

CHHANDAS

PRITHIVI
Earth

The Expansion of the Tree of Life is Located in the Constitution of the Universe

Within each of the ten Sefiroth there is said to be a complete Tree of Life. Although this idea is not commonly known, one of the early scrolls of the Kabbalistic Tree of Life graphically portrays smaller Trees of Life sprouting from the primary Tree of Life (Halevi, 1979, pp. 72-73). Hence, the expansion of the Tree of Life in such a way would create eight new octaves of Sefiroth from each of the Sefiroth in the Master Octave of the primary tree. Together this would create a total of sixty-four new Sefiroth. We propose that this possible expansion of the Tree of Life corresponds to the final stage of elaboration of the Constitution of the Universe in which the eight syllables of the first pada of Rig Veda expand into sixty-four units. Moreover, each of the sixty-four Sefiroth in the expanded Trees of Life can be viewed with respect to the triad of Da'at, Tiferet and Yesod (Holy Trinity), thus comprising a total of 192 units. Likewise in the final level of the Constitution of the Universe each of the sixty-four syllables of Rig Veda is interpreted with respect to rishi, devata and chhandas (a total of 192 units).

In summary, we have provided intriguing evidence that the structure of the Kabbalistic Tree of Life matches the numerical structure of unfoldment of the Constitution of the Universe. Perhaps the correspondence presented in this section will provide an opportunity for Hebrew scholars to revive the original or innermost teachings of the Kabbalah that have remained obscure or otherwise have been lost over the centuries.

SECTION IV

ANCIENT CHINA – THE I CHING

At the foundation of Chinese philosophy is the premise that the world is a cosmos, not a chaos, and that the ultimate frame of reference for all changes is nonchange. These ideas are the basis of one of the most pivotal books of all Chinese literature, the I Ching. Translated as "The Book of Changes," the I Ching has been a wellspring of wisdom for thousands of years, and its origin recedes into the mists of prehistoric China. Yet its influence still saturates everyday life, culture and thought in China. In addition, it has occupied the attention of the most eminent scholars of China up to the present day. Further, both branches of China's major philosophies, Taoism and Confucianism, find their origins in the I Ching. In this section we will discuss the foundational principles of the I Ching and demonstrate that there is a strikingly close parallel with the numerical structure of the Constitution of the Universe.

Origin and Purpose of the I Ching

Among the countless translations and commentaries on the I Ching, the work of Richard Wilhelm has been generally accepted to be an authoritative account of the subject. According to his well-known meticulous translation of, and lectures on, the I Ching, four men are cited in the Chinese literature to be authors of the I Ching (Wilhelm and Baynes, 1967). These four are as follows: Fu Hsi, King Wen, the Duke of Chou and Confucius. Of these the legendary figure, Fu Hsi, is thought to be the inventor of the images or symbols which compose the I Ching. Over the centuries a dense overgrowth of interpretations and extraneous ideas in understanding the I Ching has accumulated. For some sections of the I Ching there are simply no explanations and it is often difficult to comprehend what the original writers intended to convey to the readers. Nevertheless, there are a few basic concepts which appear to be unambiguous. One of the central ideas of the I Ching revolves around the idea of change. The ancient Chinese realized that change was a constant phenomenon, always happening without stop, yet out of this seeming chaos of life the I Ching creates order and describes certain immutable laws which determine the flow of events. Complete understanding of such laws will give knowledge and wisdom of both past and future events. Such ideas were founded on the premise that the microcosm and macrocosm are not separated by any fixed barriers, but are governed by the same universal laws.

Another central function of the I Ching, which is found in the earliest records of the text, is its use as a diviner's manual where the various images and symbols of the text are used as oracles (something which foretells the future). Undoubtedly the quality of the I Ching to provide knowledge of the future would have led to its use for divination. The ancient method of using the I Ching as an oracle was based on the tossing of yarrow stalks. More recently this has been replaced by tossing coins or dice.

The Active Principle of Wu Chi, the Tao, Creates Yin and Yang

According to the I Ching, there is an ultimate reality called "Wu Chi" underlying and unifying all objects and phenomena. It is considered the basis of all existence and is often described by the following words: the supreme ultimate, the original emptiness, the purest original force and the nameless one (Chia and Chia, 1989). Symbolized by an empty circle, Wu Chi represents the unmanifest which is beyond the manifest.

Another common term used by the Chinese is Tao. The Tao (pronounced as "dow") is translated as meaning "the Way" and is considered to represent the active principle or internal dynamics inherent in Wu Chi (personal communication, Douglass White). In the words of famous Lu Tzu, "That which exists through itself is called the Way (Tao). Tao has neither name nor shape. It is the one essence, the one primal spirit. Essence and life cannot be seen." (Wilhelm and Jung, 1962). The basic idea is that the Tao, although itself motionless, is the source of all movement and ways of nature. It is also commonly described as the one primal law, the essence of nature and the indefinable reality.

One of the chief goals of the ancient Chinese sages was to become one with the Tao or to live in harmony with the order of the natural world. They believed that there was no better way in which to cultivate human nature and life than to bring both back into unity with the Tao. This is well exemplified in the ancient Chinese adage, "Everything in accordance with Tao" (Rossback, 1987).

The manifestation of creation from the Tao begins first with the emergence of two polar opposites known as yin and yang (Wilhelm and Jung, 1962). From the unity and wholeness of the Tao emerges the duality of these two primal forces. The dynamic interplay of yin and yang represents the movements of the Tao which govern all aspects of change in the universe. All change and energy were observed as moving between the extremes of yin and yang. Although yin and yang are polar opposites, one cannot exist without the other. Both forces are necessary and neither one is better than the other. All aspects of creation may be classified with respect to yin and yang. For example, qualities often ascribed to yin include the following: darkness, female, negative, passive, cold, soft, wet and contraction. Opposing these qualities are the respective characteristics of yang: light, male, positive, active, heat, hard, dry and expansion.

This initial movement of the Tao into yin and yang matches well the genesis of the first syllable of the Constitution of the Universe, "A-K." The first letter, "A," represents a state of complete unity or unbounded intelligence which correlates with the descriptions of the Tao. The second letter, "K," represents a point value expression of "A" corresponding to the creation of yin and yang. Together the first two letters (first syllable) of Rig Veda, "A-K," describes the collapse of all possibilities to a point value corresponding to the division of the one primal spirit of Tao into yin and yang.

The Interaction of Yin and Yang Creates the Eight Trigrams

In the I Ching yin is written as a broken horizontal line and yang as a solid, unbroken horizontal line. The intermingling or combinations of these two symbols display the secret of the I Ching. Wilhelm (Wilhelm, 1979) explained that from the One (Tao) are created two opposites (yin and yang). The One is represented by a line, yin by a divided line and yang by an undivided line. These three form a triad which is conceived as the basis of reality. A similar description is also found in the Tao Te Ching, "Tao gave birth to the One, One gave birth to Two, Two gave birth to Three, Three gave birth to all the myriad things" (Lao, 1961, Chapter 42, p. 61).

This idea of a fundamental triad was used to define the possible combinations and interactions between yin and yang. The single lines of yin and yang were grouped into sets of three lines. Each set was called a Pa Kua or trigram ("tri" meaning three and "gram" meaning something written) and are read from the bottom up. The total number of trigrams or Pa Kua is thus exactly eight. Mathematically this is simple to determine. If we have a set of three lines and each line could be either broken or unbroken, the number of unique combinations will be eight ($2^3 = 8$). The eight trigrams represent all possible cosmic and human situations. These situations were not seen as static entities, but rather as stages of constant change and transition. As described earlier, the idea of change lies at the heart of the I Ching. Based on current information, the earliest known record of the eight trigrams is in the I Ching of the Hsia dynasty (dated between 2205 and 1766 BC).

It is interesting to reflect for a moment on the importance of the number eight to the ancient Chinese. As we have just seen, eight is the number of trigrams constituting the foundation of all aspects of the I Ching. The Chinese also recognized eight directions and eight regions. Another example of the significance of eight is located in the story of "The Eight Immortals." This myth describes the life of eight immortal beings that are the best known figures in Chinese literature. The number eight was chosen because of its "peculiar significance, a kind of 'perfect' number" (Lai, 1972). The eight immortal beings are often found on paintings, porcelainware and other common objects.

The Eight Trigrams and the Constitution of the Universe

At this stage we are able to draw our second major correspondence between the Constitution of the Universe and the I Ching. As described earlier, the first syllable of the Constitution of the Universe, "A-K," precisely matches the emergence of yin and yang from the Tao. In the next level of the Constitution of the Universe the first syllable "A-K" is elaborated into the eight syllables or prakritis of the first pada (refer to Figure 1). Likewise from the interaction of yin and yang the eight Pa Kua or trigrams are constructed. It is exciting to note the phonetic closeness of the word "prakriti" used in the Rig Veda and the word "Pa Kua" used in the I Ching.

Further, based on the descriptions attributed to each of the eight Pa Kua in the I Ching, we have discovered a striking similarity with each of the eight prakritis of Rig Veda. The following is our proposed correlations (refer to Table 13 for summary):

- The first prakriti, according to Rig Veda, is called "Ahamkara" and is translated as ego. Our ego is described as the innermost subjective quality and is very close to the level of pure consciousness. It is the sense of "I-ness." We feel that this corresponds to the Pa Kua known as "Ch'ien," translated as purity.

- The second prakriti is called "Buddhi" and is translated as the intellect. Our intellect has the attributes of discrimination, thinking and mental activity. This internal activity matches well the Pa Kua known as "Chen" which means movement.

- The third prakriti is known as "Manas" and means mind. Common qualities associated with the mind are calmness, memory and feeling. In addition, writers often metaphorically describe the mind as embracing the attributes of a body of water or a lake. The mind is like a deep lake where thoughts, like

---

Table 13. Suggested Correspondence Between the Eight Prakritis of the Constitution of the Universe and the Eight Pa Kua (Trigrams) of the I Ching

Rig Veda:							
AK	ni	mi	le	pu	ro	hi	tam
Prakriti:							
Ahamkara	Buddhi	Manas	Akasha	Vayu	Agni	Jala	Prithivi
Translation:							
Ego	Intellect	Mind	Space	Air	Fire	Water	Earth
Meaning:							
Purity	Movement	Lake	Spatial	Wind	Fire	Water	Mountain
Name:							
Ch'ien	Chên	Tui	K'un	Sun	Li	K'an	Kên
Trigram:							

bubbles, rise up to express themselves at the surface. We propose that this third prakriti is most aptly matched with the Pa Kua known as "Tui" which is translated as lake.

- The correspondence between the last five prakritis and the remaining five Pa Kua or trigrams is unambiguous. "Akasha," the fourth prakriti, translated as space, corresponds to the Pa Kua known as "K'un" meaning spatial. The fifth prakriti, "Vayu," is translated as air. By definition air in motion is wind, which precisely matches the Pa Kua known as "Sun," translated as wind. The sixth prakriti "Agni" and the Pa Kua known as "Li" perfectly match because both words are translated as fire. The same scenario is also present between the seventh prakriti "Jala" and the Pa Kua known as "K'an," both words being translated as water. Last, the eighth prakriti, "Prithivi" is translated as earth. This corresponds to the last Pa Kua, known as "Ken," translated as meaning mountain, one of the most prominent features associated with Earth.

The next level in the expansion of the Constitution of the Universe is the interpretation of the eight prakritis with respect to rishi (knower), devata (process of knowing) and chhandas (known). This produces a total of twenty-four units. Likewise, the exact same structure is present in the I Ching. In the Official Book of the Kau dynasty, which has the earliest known mention of the I Ching, three Systems of Changes are described. In each system there are rules for eight primary figures and their expansion to sixty-four. In the words of the Official Book of the Kau dynasty, "the Grand Diviner ... had charge of the rules for the three systems of Changes, called the Lien-shan, the Kwei-chang, and the I of Kau; that in each of them the primary lineal figures were eight, which were multiplied, in each, till they amounted to 64" (Legge, 1963, pp. 3-5). Unfortunately, only the I of Kau survived, which forms the present text of the I Ching. According to Legge (Legge, 1963), no trace remains of what was contained in the other two systems. Thus, it is difficult to propose how the systems of Changes precisely matches rishi, devata and chhandas. Even so, the correspondence is strong between the three systems of Changes and the flavors of rishi, devata and chhandas. We may, therefore, view the eight trigrams of Pa Kua with respect to these three systems of Changes, giving a total of twenty-four units, corresponding to the twenty-four units in the Constitution of the Universe.

Construction of the Sixty-Four Hexagrams and the Constitution of the Universe

In order to achieve greater diversity and multiplicity, the I Ching combines the eight trigrams with each other to create sixty-four new images. Each of these new images or symbols is called a hexagram and is constructed by placing one trigram above another, creating a total of six lines where each line is either broken (representing yin) or unbroken (representing yang). According to general tradition, the present collection of sixty-four hexagrams originated with King Wen, progenitor of the Chou dynasty. Following this work, his son, the Duke of Chou, added the text pertaining to the individual lines of the hexagrams. These hexagrams are the basis of the I Ching and are believed to be a microcosm representing sixty-four fundamental processes of change responsible for all possible permutations of the universe. As with the trigrams, the hexagrams can be interpreted as being related to the variety of human and cosmic situations.

There are many different ways the sixty-four hexagrams can been arranged. In Table 14 we have presented the sixty-four hexagrams in an order derived from our proposed sequence of the eight trigrams corresponding to the eight prakritis in Rig Veda (see Table 13). Using this order, we can systematically proceed to construct the sixty-four hexagrams by placing one trigram above another. For example, construction of the first line of eight hexagrams is accomplished in two steps. First, the first three lines (remembering that hexagrams are read from bottom up) of each of these hexagrams is the trigram Ch'ien. In the second step the remaining three lines in each of these hexagrams are the trigrams in their order from Ch'ien to Ken. Similarly the order and arrangement of all other hexagrams are determined. Refer to Table 14 if clarification is needed.

Interestingly, Fu hsi, the legendary author of the earliest known arrangement of the I Ching diagrams (Legge, 1963, Plate II, Fig. 1), expanded the eight trigrams into an eight-by-eight square of hexagrams using the same procedure we have used for summarizing the eight trigrams and sixty-four hexagrams in Table 14. More than one thousand years later King Wan rearranged the hexagrams in the order that we find them today in the I Ching. This arrangement is depicted in Table 14 by the numbers below each hexagram that refer to the order that they appear in the I Ching according to King Wan. Yet another arrangement of the hexagrams, presented by Wilhelm and Baynes (Wilhelm and Baynes, 1967), assigns

Table 14. Locating the Constitution
of the Universe in the I Ching

The Eight Pa Kua or Trigrams:

Ch'ien	Chên	Tui	K'un	Sun	Li	K'an	Kên

The Elaborated Sixty–Four Hexagrams:

Ch'ien	Ta Chuang	Kuai	T'ai	Hsia Ch'u	Ta Yu	Hsii	Ta Ch'u
1	34	43	11	9	14	5	26

Wu Wang	Chên	Sui	Fu	I	Shih Ho	Chun	I
25	51	17	24	42	21	3	27

Lii	Kuei Mei	Tui	Lin	Chung Fu	K'uei	Chieh	Sun
10	54	58	19	61	38	60	41

P'i	Yii	Ts'ui	K'un	Kuan	Chin	Pi	Po
12	16	45	2	20	35	8	23

Kou	Hêng	Ta Kuo	Shêng	Sun	Ting	Ching	Ku
44	32	28	46	57	50	48	18

T'ung Jên	Fêng	Ko	Ming I	Chia Jên	Li	Chi Chi	Pi
13	55	49	36	37	30	63	22

Sung	Hsieh	K'un	Shih	Huan	Wei Chi	K'an	Meng
6	40	47	7	59	64	29	4

Tun	Hsiao Kuo	Hsien	Ch'ien	Chien	Lii	Chien	Kên
33	62	31	15	53	56	39	52

Arrangement according to the suggested correspondence between
the eight Pa Kua and the eight prakritis *(see Table 13)*. Numbers below each
Hexagram refer to the order that they appear in the I Ching by King Wan.

the sixty-four hexagrams to eight houses. No one arrangement or order of writing the hexagrams appears to be correct because the different ways have served various functions of the I Ching throughout the centuries.

The construction of the sixty-four hexagrams from the eight trigrams provides a complete and striking match with the Constitution of the Universe. In the final stage of elaboration of the Constitution of the Universe the eight syllables of the first pada of Rig Veda expand into sixty-four units. These sixty-four units exactly match the sixty-four hexagrams. Moreover, each of the sixty-four hexagrams can be viewed with respect to the three Systems of Changes (the Lien-shan, the Kwei-chang and the I of Kau), thereby totaling 192 units. Likewise in the final level of the Constitution of the Universe each of the sixty-four syllables in Rig Veda is interpreted with respect to rishi, devata and chhandas (a total of 192 units).

In summary, we have discovered that the foundational principles of the I Ching correspond exactly with the Constitution of the Universe. This correspondence is particularly strong considering that it is based on the numerical structure of the I Ching which has remained invariant throughout the ages as opposed to the commentaries and other qualitative descriptions. It is also interesting that among the classics of ancient China the I Ching is one of the few books that survived the historic burning of Confucius' library under the tyrant Ch'in Shih Huang Ti and the vicissitudes of the centuries.

As an ending note it is worth mentioning that a number of other independent authors have also found striking parallels between the I Ching and universal principles. As mentioned earlier (see Chapter 4, Section I), at least three individuals have published evidence on the association between the I Ching and the DNA: Gunther Stent (Stent, 1969), Dr. Schonberger in 1973 (translated into English, Schonberger, 1992), and Johnson Yan (Yan, 1991). Another fascinating relationship was observed by an author Jose Arguelles who located in the Tzolkin or ancient Mayan calendar system the structure of the I Ching (Arguelles, 1984 and 1987). These published links between the I Ching and the DNA and the Mayan calendar not only add further significance to ancient Chinese wisdom, but also naturally suggest that there should be a parallel between the DNA and Mayan Calendar and the structure of the Constitution of the Universe. Such links do, in fact, exist and are presented in Chapter 4, Section I and Chapter 8, Section VII, respectively.

SECTION V

ANCIENT EGYPT – THE RIGHT EYE
OF HORUS AND THE FLOWER OF LIFE

"I am One that transforms into Two
I am Two that transforms into Four
I am Four that transforms into Eight
After this I am One"

The Ancient Hermopolitan Mystery,
Coffin of Petamon, Cairo Museum no. 1,160

Throughout the ancient world the Egyptian mystery schools and temples were acclaimed for their wealth of knowledge. For several thousand years Egypt remained a world power and thrived on the advanced wisdom of life. Not only did they have remarkable architectural achievements, such as the famous pyramids, but they also possessed a deep spiritual understanding of nature. Over time this wisdom has been lost and what we have remaining today are countless hieroglyphic inscriptions and mystical remnants. Although many individuals have written on the topic, the wisdom of ancient Egypt still remains an enigma. Recently, however, several authors have provided profound insight into various aspects of ancient Egyptian wisdom. The information in this section is primarily based on the work of Drunvalo Melchizedek who reportedly obtained his knowledge through direct communication with Thoth, one of the ancient Egyptian gods. In addition to summarizing this work, we also demonstrate that this wisdom is identical to the structure of the Constitution of the Universe.

Sacred Geometry in the
Egyptian Mystery Schools

In ancient Egypt during the period of the pharaoh Akhunaton there were two schools of learning (Frissell, 1994). The first Egyptian mystery school was known as the "Left Eye of Horus" and comprised twelve years of training. During the course of study in this school the initiates were taught experiential techniques for the development of many different aspects of human nature. Included in these were programs to deal directly with fears and emotions, which enabled the students to integrate these different aspects into their greater sense of well-being.

Based on successful completion of the "Left Eye of Horus," the student spent an additional twelve years in the second Egyptian Mystery School in what was known as the "Right Eye of Horus" or the "Law of One." This second school gave the initiate the complementary knowledge to their experiential understanding, that is, the intellectual understanding of the universe. This program, based on the knowledge of Sacred Geometry, was considered to be one of the most important aspects of Egyptian wisdom.

Sacred Geometry is the logical knowledge of how everything in the universe was created by the Spirit alone. It is the language of how forms and shapes give rise to the structure of the ultimate reality. Often Sacred Geometry is referred to as the "language of light" or the "language of silence" from which the entire universe continues to unfold in perfect order. The only place left in Egypt where all the images of Sacred Geometry are displayed is on the upper part of the left-hand side of a long hallway, underneath the Great Pyramid, leading into what is known as the Hall of Records (Frissell, 1994). For the most part, the secret teachings of the Egyptian mystery schools were transmitted orally; this explains why this knowledge has been missed by virtually all historians and Western scholars.

The Flower of Life is a Symbol Containing All Knowledge

One of the most important images in Egyptian Sacred Geometry is the Flower of Life. Although this image is found in Egypt, it is also located throughout the ancient world in countries such as India, Tibet, China, Scotland, Ireland and Turkey. Hidden in the image of the Flower of Life is the blueprint of everything in the universe, of all aspects of knowledge. It appears that, except for the work of Drunvalo Melchizedek and the knowledge imparted to him from the ancient Egyptian god, Thoth (Melchizedek, 1992; Frissell, 1994), very little research has been done on the Flower of Life image. Our purpose in this section will be to focus on the teachings of the "Right Eye of Horus" as displayed in the image of the Flower of Life.

According to the ancient Egyptians (Melchizedek, 1992; Frissell, 1994), the mechanics of creation, as symbolized in the construction of the Flower of Life, originates from a state of voidness or total nothingness. Voidness represents "the Beginning" where consciousness or the spirit resides. All knowledge necessary to form anything is present in the spirit or in this level of consciousness. From this voidness the universe is created in a sequential pattern and is described in Sacred Geometry using shapes and forms. Although our discussion will revolve around drawing lines on paper, what we are actually representing is the movement of consciousness in the Void.

Creation of the Octahedron and Sphere from the Field of Voidness

Prior to the creation of anything, the spirit alone exists. As long as the spirit remains alone in the Void, there is no point of reference. Without another object in space and time to act as a reference point there is no concept of motion and thus movement remains impossible. Hence the first stage of creation is to construct a form making movement possible. Inherent in the nature of consciousness is the quality of three-ness which becomes expressed as three dimensional space. In creating a three dimensional arena, six rays emerge: front and back, left and right, and up and down (refer to Figure 28). The next stage is to create an eight-sided octahedron from these six rays. To construct an octahedron, we first need to create a square. This becomes possible if we simply connect the ends of the back, left, front and right rays. Using this square as the base, two pyramids can be formed, back to back, one above and one below the square. The apex of the first pyramid is the end of the "up" ray and the apex of the second pyramid is the end of the "down" ray. The construction of these two "back to back" pyramids completes the construction of the octahedron which is visually represented in Figure 28.

This octahedron is the first image created around consciousness. With this physical object movement becomes possible. Rotation about the three axes of the octahedron (x, y and z) traces out the image of a sphere which is one of the most important shapes. Up until the creation of this sphere we have been dealing with straight lines only. However, in creating the sphere we have our first appearance of curved lines. From Sacred Geometry, straight lines represent "male" and curved lines delineate "female." According to Melchizedek, this represents the first of many

Figure 28. Mechanics of Creation and Construction of the Flower of Life According to Sacred Geometry

Stage One: Mechanics of Creation

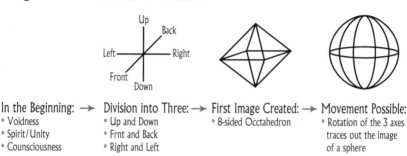

In the Beginning: →
- Voidness
- Spirit/Unity
- Counsciousness

Division into Three: →
- Up and Down
- Frnt and Back
- Right and Left

First Image Created: →
- 8-sided Occtahedron

Movement Possible:
- Rotation of the 3 axes traces out the image of a sphere

Stage Two: Cronstruction of the Flower of Life

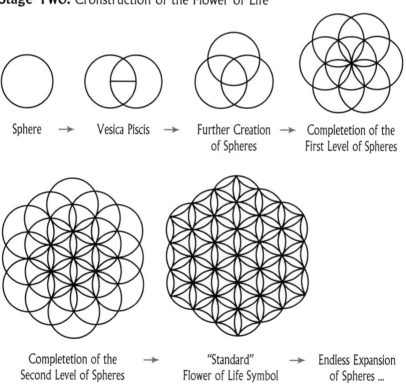

Sphere → Vesica Piscis → Further Creation of Spheres → Completetion of the First Level of Spheres

Completetion of the Second Level of Spheres → "Standard" Flower of Life Symbol → Endless Expansion of Spheres ...

Adapted from Frissell, B. (1994). *Nothing in This Book Is True, But It's Exactly How Things Are.* Berkeley, California: North Atlantic Books.

connections with Genesis in the Holy Bible. The Bible states, "God then made the man fall into a deep state of unconsciousness, and he slept. He took one of his ribs and closed the flesh in its place. God built the rib that he took from the man into a woman, and He brought her to the man" (Kaplan, 1981, p. 8). Similarly, in the construction of the Flower of Life straight lines (male or Adam's rib) were the first shapes to be created from the Void. From these straight lines the curved lines of the sphere (female or Eve) were constructed.

Thus far we have constructed a sphere out of a field of voidness. This initial development of the Flower of Life symbol matches creation of the first syllable of the Constitution of the Universe, "A-K." The first letter, "A," represents a state of complete unity and field of all possibilities corresponding exactly to the idea of consciousness existing in a field of voidness. The second letter, "K," represents a point value which would correspond to the creation of a bounded structure, in this case represented by the sphere. Together the first two letters (first syllable) of Rig Veda, "A-K," describe the collapse of fullness to a point value. This collapse takes place in a sequence of *eight* steps subsequently giving rise to the first eight syllables of Rig Veda making up the first pada (refer to Figure 1). Intriguingly this same pattern is found in the construction of the Flower of Life symbol. From a state of voidness ("A"), consciousness constructs the image of a sphere ("K") through the rotation of an *eight*-faced octahedron (eight-fold collapse of voidness to space-time boundaries).

The First Motion of Consciousness Creates the Vesica Piscis

Construction of the sphere allows for the first motion of consciousness. However, the only thing that consciousness can do at this stage is to move to any point on the surface of the sphere. According to Melchizedek, this corresponds to the sentence in Genesis which states, "In the beginning God created heaven and earth. The earth was without form and empty, with darkness on the face of the depths, but God's spirit moved on the water's surface" (Kaplan, 1981, p. 3). The movement of the Spirit to the "water's surface" corresponds to the movement of consciousness to the surface of the sphere. After this first movement everything else to be created in the entire universe is automatic.

Once at the surface of the sphere, consciousness simply creates another sphere. The image of these two spheres is called the "vesica piscis." These spheres do not overlap one another, but do touch at their surfaces. For simplicity we will represent these two spheres, as well as each subsequent stage in the construction of the Flower of Life, in two dimensions. We should remember, however, that in reality all the circles we will be constructing do in fact represent three-dimensional spheres. The two-dimensional image for the vesica piscis is shown in Figure 28. In this figure both circles are of identical size, and a straight line, the two-dimensional distance of one radius, is shown connecting the centers of both circles. Hence, the second circle overlaps the first circle by the distance of one radius. (Note: as stated above, in three dimensions there is no actual overlap of the two spheres. The overlap of the two circles in two dimensions depicted in Figure 28 arises from viewing the three dimensional transparent spheres in a two dimensional arena. In three dimensions the second sphere is *behind and slightly to the right* of the first sphere in such a way that when viewed in two dimensions there appear to be two circles overlapping by the distance of one radius.)

The symbol of the vesica piscis is found throughout the world, representing one of the most basic symbols of creation. For example, the vesica piscis is the main image on the lid of the chalice well in Glastonbury, England. In addition, the first proof in Euclidean Geometry uses the image of the vesica piscis to construct an equilateral triangle (Todhunter, 1948). (The three points forming the equilateral triangle are the centers of the two circles and either one of the points where the two circles intersect.) Construction of this second sphere or circle corresponds to the first day of creation. According to Genesis, on the first day of Creation, the Bible states, "God said, 'There shall be light', and light came into existence" (Kaplan, 1981, p. 3). This excerpt matches the image of vesica piscis representing the morphogenic structure of the eye which is responsible for vision and hence the perception of light (Frissell, 1994).

Construction of the Complete Image of the Flower of Life

The next stage in the construction of the Flower of Life symbol arises from the creation of more spheres or circles. In two dimensions, each new circle drawn is placed at the distance of one radius from any of the surrounding circles. It turns out that five additional circles are needed to make one complete "level" around the primary sphere, that is, exactly six spheres in all (refer to diagram labeled "Completion of the First Level of Spheres" in Figure 28). According to Melchizedek, each of these new circles corresponds to another day in Genesis. The six circles needed to complete the first level exactly correspond to the first six days of Genesis needed to create both Heaven and Earth. At this stage in the construction of the Flower of Life symbol, no additional unique circles can be constructed on the first level. To draw a seventh circle is redundant because we would simply trace the image of one of the pre-existing circles. Similarly, the Bible states, "With the seventh day, God finished all the work that He had done. He [thus] ceased on the seventh day from all the work He had been doing. God blessed the seventh day, and He declared it to be holy" (Kaplan, 1981, p. 5). There was nothing more that needed to be done on the seventh day (or seventh circle) because all the laws in the universe had been created.

Further expansion of the Flower of Life symbol proceeds through the construction of second-level circles, using the same method described above. The completion of the second level of spheres around the primary sphere is depicted in Figure 28. Twelve circles are needed to construct this second level as distinct from the six circles needed in the construction of the first level. In this way, an endless expansion of additional levels of circles can be constructed in the Flower of Life symbol. However, the "standard" Flower of Life symbol which is usually depicted in the ancient civilizations is displayed in Figure 28. Hidden within this "standard" Flower of Life symbol are numerous shapes and structures describing the universe. The trick to locating these structures and shapes is to erase particular lines while emphasizing others.

One of the most important images hidden in this Flower of Life symbol is a cluster of eight spheres forming the heart of the Flower of Life diagram (refer to Figure 29). The eighth sphere is hidden behind the seven which are visible. To the Ancients this image was known as the "Egg of Life" and represents one of the most important patterns. In the words of Melchizedek, the Egg of Life is a "Pattern which forms all biological life forms ... a pattern of ALL structure – no exceptions whatsoever" (Melchizedek, 1992). This image is found in the harmonics of music and the harmonics of the electromagnetic spectrum and is a pattern of all biological life forms.

Figure 29. Locating the "Egg of Life" in the Flower of Life

Adapted from Frissell, B. (1994). *Nothing in This Book Is True, But It's Exactly How Things Are.* Berkeley, California: North Atlantic Books.

The Structure of the Flower of Life is Identical to the Constitution of the Universe

At this stage we can draw our complete correspondence between the Flower of Life symbol and the Constitution of the Universe. The first level of unfoldment of the Constitution of the Universe is from the first syllable of Rig Veda, "A-K," to form the eight syllables of the first pada (refer to Figure 1). As described earlier, the first syllable, "A-K," precisely matches the construction of the primary sphere from the Void. We further propose that the eight fundamental syllables of Rig Veda arising from "A-K" correspond to the eight spheres forming the Egg of Life. Not only is this image located in the center of the Flower of Life symbol and is constructed from the primary sphere, but it was also recognized by the Ancients to be one of the most important images in the universe.

From the eight basic syllables of the first pada of Rig Veda, the next level of expansion in the Constitution of the Universe is the interpretation of each of these eight syllables with respect to rishi (knower), devata (process of knowing) and chhandas (known). This produces a total of twenty-four units. Likewise in the Flower of Life symbol, as described earlier, consciousness by nature has the structure of three-ness, expressed as three dimensional space. Each of the three dimensions corresponds to the three aspects of rishi, devata and chhandas. The first dimension is the most abstract and is represented by a straight line or linear information corresponding to the abstract nature of rishi. The second dimension, representing the concept of area, acts as a transitional dimension between the first dimension and the third dimension. This corresponds nicely with the qualities embodied by the devata interpretation which is to connect or act as a transitional state. Last, the third dimension represents the most concrete expression of the three dimensions and is thus aptly equated with the material nature of the chhandas interpretation. These three dimensions, when viewed with respect to each of the eight spheres in the Egg of Life, produce a total of twenty-four units corresponding to the twenty-four units in the Constitution of the Universe.

In the final stage of elaboration of the Constitution of the Universe the eight syllables of the first pada of Rig Veda expand into sixty-four units. A similar structure can also be located in the Flower of Life symbol. As described earlier, the Flower of Life symbol can be viewed as an endless expansion of levels of spheres beyond what is represented in the "standard" symbol. Hence in addition to locating the basic cluster of eight spheres that make up the Egg of Life, it would also be easy to locate a central cluster of sixty-four spheres in a slightly more elaborate diagram of the Flower of Life. We propose that these sixty-four spheres correspond to the elaborated sixty-four syllables in the Constitution of the Universe. Further, each of these sixty-four spheres can be viewed with respect to the three dimensions, thus comprising a total of 192 units. Likewise in the final level of the Constitution of the Universe each of the sixty-four syllables of Rig Veda are interpreted with respect to rishi, devata and chhandas (a total of 192 units).

The Kabbalah and Platonic Solids are also Hidden in the Flower of Life

In concluding this section, we would like to provide an additional glimpse into the importance of the Flower of Life symbol by locating some other

Figure 30. Locating the Structure of
the Kabbalah in the Flower of Life

Adapted from Frissell, B. (1994). *Nothing in This Book Is True, But It's Exactly
How Things Are.* Berkeley, California: North Atlantic Books.

important structures in it. In particular we will draw further upon the work
of Melchizedek (Melchizedek, 1992; Frissell, 1994) to find the hidden
structures of the Kabbalah and five Platonic Solids in the "standard" Flower
of Life diagram. Both these examples were chosen because the significance
of these symbols are discussed elsewhere in this book.

Figure 30 displays the symbol of the Kabbalah (also known as the Tree of
Life) located in the Flower of Life. In Chapter 8, Section III we describe in
detail what the Kabbalah is and how this symbol also embodies the
numerical structure of the Constitution of the Universe. Locating the
Kabbalah symbol here provides confirmation of matching the Flower of
Life symbol with the Constitution of the Universe.

Figure 31. Locating the Five Platonic
Solids in the Flower of Life

Dodecahedron
12 faces, 30 edges, 20 corners

Octahedron
8 faces, 12 edges, 6 corners

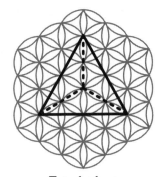

Tetrahedron
4 faces, 6 edges, 4 corners

Icosahedron
20 faces, 30 edges, 12 corners

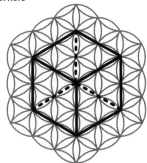

Cube
6 faces, 12 edges, 8 corners

Adapted from Frissell, B. (1994). *Nothing in This Book Is True, But It's Exactly
How Things Are.* Berkeley, California: North Atlantic Books.

Figure 31 demonstrates how each of the five Platonic Solids are hidden in the Flower of Life diagram. These five regular polygons were first described by Plato (hence "Platonic Solids") to represent five fundamental forces responsible for the structure of the universe. These five solids are: the cube, the tetrahedron, the dodecahedron, the octahedron and the icosahedron. By definition, each of these solids must satisfy four conditions: first, all edges on the solid are equal; second, all angles on the solid are the same; third, all the faces on the solid are identical; and fourth, when placed in a sphere all the points of the solid must touch the surface of the sphere. These five solids are the only polygons which meet all four criteria. From these five solids all other geometric shapes are created. We present this last connection to demonstrate the mechanics of how from the sphere of voidness the Flower of Life symbol emerges which in turn provides the framework for creating the Five Platonic Solids. It is important to recognize that all five Platonic Solids can be constructed from the sphere. Understanding this concept provides the foundation for Chapter 16 of this book, in which we go into a more detailed discussion of the five Platonic Solids and their significance in understanding the genesis of the material universe.

Section VI

Ancient Greece – The Sacred Tree Script in Plato's Nuptial Number

"Know thy Self."

– Socrates

"And the ancients, who were superior to us and dwelt nearer to the Gods, have handed down a tradition that all things that are said to exist consist of a ONE and a Many and contain in themselves the concrete principles of Limit and Unlimitedness."

– Plato
(Efron, 1941, p. 321)

For over 2,500 years Platonic wisdom from ancient Greece has given both direction and depth to numerous disciplines of the Western world. Although the source of Plato's knowledge is not fully known, it seems that it may have been derived from both his Pythagorean schooling and his famous teacher Socrates. During his life Plato founded the Platonic Academy in Athens and preserved his teachings in numerous books and writings such as the *Republic*, *Symposium*, *Phaedo*, *Sophist* and *Theaetetus*. However, even though these records have survived for centuries, the real truth and meaning of many of the central ideas of his wisdom are still obscure. In this section we present evidence, primarily derived from the work of Andrew Efron (Efron, 1941), of the existence of a universal template known as the "Sacred Tree Script," which provides a unique way of understanding a foundational aspect of Plato's wisdom. In addition to summarizing this work, we also demonstrate that both the structure of this "Sacred Tree Script" and the work of Plato match the structure of the Constitution of the Universe. We have divided this section into two parts. In the first part we describe the "Sacred Tree Script" and its relation to the

Constitution of the Universe. Based on this analysis, the second part locates within the structure of the "Sacred Tree Script" a key aspect of Platonic knowledge.

The Sacred Tree Script

The "Sacred Tree Script" is essentially a fir tree symbol (refer to Figure 32) in which the number of "twigs" on each side, their thickness and their length are allocated universal significance. For example, the number of twigs on both sides are unequal, the twigs on the right side differ in length from those on the left side, and the upper part of the tree symbol differs from the lower part. There is only one known location where this symbol still exists, on the Runic Stone of Kylfver. Although this symbol is not directly described in any of the ancient sources, Efron claims that indirect evidence supports the notion that the "Sacred Tree Script" was a chief symbol of ancient science and that contemplation of this symbol may have given rise to the different branches of science (Efron, 1941).

The symbol of a fir tree or evergreen tree represents an expression of natural order which provides a medium between the external world and the human mind. Such symbols have occupied an important place in many cultures (e.g., the Finns, Germans, Slavs, Anglo-Saxons and Norsemen). The symmetry and regularity of the tree symbol lends itself well to its use as a counting table, a textbook on arithmetic, alphabets, ethics and a primitive encyclopedia. As nicely put by Efron, the "tree symbol might well be called a 'synopsis' of ancient science: everything had to be reflected in it and almost everything actually had been derived from it" (Efron, 1941, p. 134) and, "it was more than a mere symbol, a 'human convention' – it was knowledge itself, 'a divine gift from heaven to mankind' – the central point of the esoteric ancient wisdom" (Efron, 1941, p. 64).

The inherent structure of the "Sacred Tree Script" is based on an "octaval" system that provides the key to its understanding and application. In this symbol the octaval system is derived from the eight twigs on the right hand side (refer to Figure 32). From this octaval structure of eight twigs all manifestations of order and intelligence of ancient wisdom are created (Efron, 1941). Two brief examples will serve to illustrate this point. First, the "Sacred Tree Script" can be interpreted as a musical scale in which the proportions of the twigs correspond to the intervals of music. For

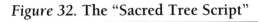

Figure 32. The "Sacred Tree Script"

Adapted from Efron, A. (1941). *The Sacred Tree Script: The Esoteric Foundation of Plato's Wisdom.*
New Haven, Connecticut: The Tuttle, Morehouse and Taylor Company.

instance, the musical octave is represented by the eight twigs on the right
side of the symbol. The musical fourth (frequency ratio of 4/3) is represented by the four twigs on the top right spanning the same length as
three twigs on the top left. The musical fifth (frequency ratio of 3/2) is
given by twigs four through six on the right side matching twigs three and
four on the left side. Thus it is likely, according to Efron, that the musical
scale was octaval from the beginning. The second example is the Gothic
alphabet which by nature is octaval. The twigs on the left side of the tree
correspond to three "families" in the Gothic alphabet. That is, the first
three twigs correspond to the first family, the next two twigs to the second
family, and the last twig to the third family. There are eight letters for each
family which derive from the right hand side of the "Sacred Tree Script"
(three families multiplied by eight letters per family gives twenty-four total
letters to the Gothic alphabet).

Locating the Constitution of the Universe in the "Sacred Tree Script"

Our first comparison between the "Sacred Tree Script" and the
Constitution of the Universe, as displayed in Rig Veda, begins with "A-K."

According to the Vedic Literature, "A" represents complete knowledge which condenses to the point value of "K." The letter "K" reflects a state of zero possibilities or specific information. Correspondingly, all wisdom of life ("A") can be represented in the concrete symbol ("K") of the "Sacred Tree Script." The Constitution of the Universe subsequently unfolds the first syllable "A-K" into the eight syllables or prakritis of the first pada (refer to Figure 1). Likewise, in the "Sacred Tree Script" the eight twigs located on the right side symbolize its inherent octaval nature.

From the eight syllables of the first pada of Rig Veda, the next level of expansion in the Constitution of the Universe is the interpretation of each of these eight syllables with respect to rishi (knower), devata (process of knowing) and chhandas (known). This produces a total of twenty-four units. Likewise, inherent within the structure of the "Sacred Tree Script" are three different counting systems (Efron, 1941). Each counting system may be applied to any twig or aspect of the "Sacred Tree Script" symbol. If we focus on the octaval nature of the tree symbol (i.e., the eight twigs on the right side), and apply each of the three counting systems to each twig, we will obtain twenty-four variations. The first counting system of the "Sacred Tree Script" is the octaval counting system (based on the number eight). This system provides the key to understanding the "Sacred Tree Script" and also provides the foundation for the other two counting systems. In terms of the Constitution of the Universe this counting system corresponds to the abstract and profound nature of rishi. The second counting system of the "Sacred Tree Script" is the Babylonian sexagesimal system which is based on the number sixty and appears to have cosmological significance. This system is derived from the octaval system and matches the nature of devata which is to provide a link between rishi and chhandas. The final counting system inherent in the structure of the "Sacred Tree Script" is the decimal system (based on the number ten). Numbers expressed in the decimal system are easily conceptualized and resemble the concrete qualities embodied by the chhandas interpretation of Rig Veda.

It appears that the current symbol of the "Sacred Tree Script" was modified or derived from an earlier "enlarged octaval system" (Efron, 1941, pp. 58-60). According to Efron, "the end form of all signs of the tree symbol suggests the idea of an initial form in which both sides of the tree had an equal number of twigs – eight twigs on each side" (Efron, 1941, p. 151). This would simply require the addition of two more twigs to the left hand side of the current tree. Eight twigs per side represents a state of "perfect-

Figure 33. Locating the Constitution of the Universe in the "Enlarged Octaval System" of the "Sacred Tree Script"

$$64 + 36 = 100$$

Adapted from Efron, A. (1941). *The Sacred Tree Script: The Esoteric Foundation of Plato's Wisdom.* New Haven, Connecticut: The Tuttle, Morehouse and Taylor Company, p. 125.

ness," "completeness" and "balance." In this "enlarged" system each twig on the left side would represent eight, with a total of sixty-four when all the left twigs are added together (refer to Figure 33). Combining these with the eight twigs on the right side (each representing the number one), we would arrive at a grand total for the "Sacred Tree Script" of seventy-two. It is worth mentioning that this "enlarged octaval system" also contains the idea of a decimal system (e.g., $64 + 1 + 2 + \ldots + 8 = 100$).

This "enlarged octaval system" provides a complete match between the "Sacred Tree Script" and the Constitution of the Universe. In the final stage of elaboration of the Constitution of the Universe the eight syllables of the first pada of Rig Veda expand into sixty-four units, each of which is interpreted with respect to rishi, devata and chhandas (a total of 192 units). Likewise in the "Sacred Tree Script" the number sixty-four is obtained from the eight twigs of the left side of the tree symbol (each twig represents the number eight, hence $8 \times 8 = 64$). Further, each of these sixty-four numbers may be used in relation to the three counting systems discussed earlier (i.e., octaval, sexagesimal and decimal), which gives rise to 192 unique aspects corresponding exactly to the numerical structure of the Constitution of the Universe.

So far we have established a strong link between the numerical structure of the Constitution of the Universe and the "Sacred Tree Script" as described by Efron (Efron, 1941). In itself this is an impressive correspondence lending credence to Efron's claim that the "Sacred Tree Script" was an all-embracing symbol of the ancient world. To provide additional support for his ideas, Efron devoted a large portion of his book to presenting "indirect evidence" that the structure of the "Sacred Tree Script" was understood by the ancient Greeks and that it can be located in Plato's Nuptial Number. In the second half of this section we will summarize this intriguing research.

Plato's Nuptial Number

The Nuptial Number is:

> *"… not merely in itself a simple and elegant mathematical problem, but forms an organic and essential part of the argument of the Republic, and furnishes us with the right point of view from which to study the cosmology of the Timaeus. It is also full of interest for students of theology, as well as of ancient astronomy, embryology, and music."*

> – James Adam,
> *Adam*, 1985, p. 17

Clearly the ancient Greek philosophers and thinkers believed that numbers were the secret origin of all things. They were not so much concerned with number manipulation, but rather they believed numbers had different forms that were key to understanding the spiritual and physical universe. Such ideas were clearly expressed by Plato, who developed an elaborate "science of numbers" to explain the principles and forces that govern reality. One of the focal points of Plato's "science of numbers," the Nuptial Number, represents a perfect number that was considered to be final and complete. It describes the source of change and governs both the growth and decay of individual life, city life and the life of the cosmos. By describing both individual and cosmic life, the Nuptial Number also unifies earthly life with the ideal life or the world of "true" reality. Calculations involving the Nuptial Number have been applied to understanding and predicting events such as birth, death, marriage, life and growth.

Mathematical Derivation
of the Nuptial Number

Individuals with the proper training from the Platonic Academy could easily interpret the Nuptial Number. However, for centuries following the death of Plato the solution to his Nuptial Number remained an insoluble mystery. For this reason many translators of Plato's writings have evaded the topic. However, around the turn of the century several authors made significant leads in deciphering the descriptions given in Plato's *Republic*, which eventually culminated in its complete solution by James Adam (Adam, 1985). This solution has been accepted and recognized by most scholarly critics. In brief, we will now summarize the derivation of the Nuptial Number, leaving the interested reader to refer to Adam's work for greater detail. Our purpose in presenting this solution is to demonstrate that the particular numbers and calculations chosen by Plato in order to derive his Nuptial Number provide evidence that it was formulated based upon the knowledge of the "Sacred Tree Script." Before presenting the solution, we will first provide the reader with the mysterious and obscure passage from the eighth chapter of Plato's Republic which describes the Nuptial Number:

> "... Now for a divine creature there is a period which is comprehended by a number that is final, [A] and for a human the number is the first in which multiplications of root by square, having laid hold on three distances, with four limits, of that which maketh like and unlike and waxeth and waneth, have rendered all things conversable and rational with one another: [B] whereof the base, containing the ratio of four to three, yoked with five, furnishes two harmonies when thrice increased: [C] the one equal an equal number of times, so many times a hundred, [D] the other of equal lengths one way, the other way unequal; – on the one side, of one hundred squares rising from rational diameters of five diminished by one each, or if from irrational diameters by two; on the other, of one hundred cubes of three. The sum of these, a number measuring the earth, is lord of better and worse births, which not knowing, when your guardians marry brides to bridegrooms out of season, children of ill nature and ill fortune will be born: whereof the best their predecessors will indeed make rulers; nevertheless, being unworthy, when they have succeeded to their fathers' offices of power, us they will first begin to heed not though they

are our guardians, having set too little store by music first and second
by gymnastic, and so our children will grow up without us ...”

– Plato's Republic
Chapter VIII, 545 D to 547 A
(from *Adam*, 1985, p. 24)

For the purpose of clarity we will describe the solution to the Nuptial Number by breaking the above passage into four separate parts, [A], [B], [C] and [D]. Each part will be deciphered individually according to the research of Adam (Adam, 1985).

Part A:

“... *and for a human the number is the first in which multiplications*
of root by square, having laid hold on three distances, with four limits,
of that which maketh like and unlike and waxeth and waneth, have
rendered all things conversable and rational with one another ...”

To decipher the Nuptial Number from Plato's writings we need to translate his words into the language of mathematics. The excerpt from this first part provides the framework that will be used in Part B through D to derive the Nuptial Number. Several key phrases when understood provide the meaning for this first excerpt. The words, “*multiplications of root by square*” are understood to mean root-and-square multiplications (e.g., $y \times y^2$). Following these words is the phrase, “*having laid hold on three distances, with four limits,*” which is understood in the following terms. Let AB, BC, CD represent “three distances,” and let A, B, C and D delineate four points or “limits” confining the three distances. These three distances correspond to 3, 4 and 5, respectively, which are the numbers of the Pythagorean right-angled triangle. As we will see in Part B, the dimensions of this right-angled triangle are central to the solution of the Nuptial Number. Combining the mathematical statements for the above two phrases we obtain:

$$(X \times X^2) + (y \times y^2) + (z \times z^2)$$

Where:

if $X = 3, y = 4, z = 5,$

then $(3 \times 3^2) + (4 \times 4^2) + (5 \times 5^2) = 216.$

The remaining words of this first excerpt have no mathematical equivalent, but rather describe the importance of the number 216. It seems that this number refers to the human period of gestation and the three terms used to calculate the number signify three distances from the time of the child's conception. A deeper discussion, provided by Adam (Adam, 1985), is beyond the scope of this book. Our main emphasis lies in the Nuptial Number.

Part B:

> "… whereof the base, containing the ratio of four to three, yoked with five, furnishes two harmonies when thrice increased …"

This excerpt describes the solution to the Nuptial Number. The words, *"whereof the base, containing the ratio of four to three,"* refer to the famous Pythagorean triangle that provide a cornerstone to deciphering the Nuptial Number. The Pythagorean triangle is a right-angled triangle consisting of the following dimensions: the upright side of the triangle equal to 3, the base equal to 4, and the hypotenuse equal to five. Adding the numbers for each of these three sides gives the number 12 (3 + 4 + 5 = 12). When 12 is "yoked" or multiplied with 5 we obtain the number 60. Then, the final step, according to the excerpt, is to increase "thrice" the number 60. Although most authors have taken these words to imply "raised to the third power" or "multiplied by some three factors," Adam (Adam, 1985) argues that it refers to "raised to the fourth power" for the following reason. The words *"thrice increased"* refer to three separate *processes* of multiplication rather than three factors which are multiplied together. Therefore, 60 multiplied *once* by 60 gives 60^2; 60 multiplied *twice* by 60 produces 60^3; and 60 multiplied *thrice* by 60 would equal 60^4. Multiplying out 60^4 gives $60 \times 60 \times 60 \times 60$ which equals **12,960,000**, the solution to the Nuptial Number.

The final words of this excerpt, *"furnishes two harmonies,"* suggests however that there are two equations or "harmonies" which will also give the Nuptial Number. These two "harmonies" are presented in Parts C and D.

Part C:

> "… the one equal an equal number of times, so many times a hundred … "

This phrase describes the first "harmony" or equation to calculate the same Nuptial Number presented in Part B. Translation of these words into mathematical terms is relatively straightforward. The words *"equal an equal number of times"* means to square a number X, and the words *"so many times a hundred"* would mean to multiply this number X by 100. The question remaining is, "What is X?" Since we know what the Nuptial Number is from part B, we can solve for X by formulating an equation:

$$(100 \times X)^2 = 12{,}960{,}000$$

Where:

$$X = 36.$$

That is, $(100 \times 36)^2 = 3{,}600^2$ (the first harmony) which is equal to 12,960,000.

Part D:

> "... *the other of equal lengths one way, the other way unequal; – on the one side, of one hundred squares rising from rational diameters of five diminished by one each, or if from irrational diameters by two; on the other, of one hundred cubes of three* ... "

This excerpt describes the derivation of two numbers corresponding to the second "harmony," from which the Nuptial Number can also be calculated. The first of these numbers obtained from the last few words of the passage, *"on the other, of one hundred cubes of three"* is simply translated in mathematical terms as $100 \times 3^3 = 2{,}700$. The second number is more difficult to decipher and is hidden in the first section of the passage. To begin, we need to understand the words, *"rational diameters of five."* A rational number is a whole number (without fractions or decimal points). Hence, these words appear to be asking what is the closest rational number, X, which is equal to the diagonal (diameter) of a square whose side is five. Mathematically, this would be $X^2 = 5^2 + 5^2$; $X = \sqrt{50}$ which equals 7 (the closest rational number). Examining these words in the context of the sentence, *"on the one side, of one hundred squares rising from the rational diameters of five,"* we obtain the mathematical statement:

$$100 \times 7^2 = 4{,}900$$

Just a few more points need to be deciphered to arrive at the full meaning of this passage. The words, *"diminished by one each"* implies subtracting a quantity. Thus, the sentence, *"on the one side ... diminished by one each"* becomes translated as:

$$(7^2 \times 100) - (1 \times 100) = 4900 - 100 = 4{,}800$$

The remaining words, *"or if from irrational diameters by two,"* refers to another way to calculate the number 4,800. Instead of using the rational number 7 as equaling $\sqrt{50}$, we could use the actual or irrational number of $\sqrt{50}$. By using the irrational number $\sqrt{50}$, we are however required to minus "two" instead of "one," as stated in the excerpt. The mathematical statement for this would be:

$$[(\sqrt{50})^2 \times 100] - (2 \times 100) = 5,000 - 200 = \textbf{4,800}$$

In summary, we have arrived at two numbers **4,800** and **2,700** which represent the second "harmony." The relationship between these two numbers is expressed in the words, *"the other of equal lengths one way, the other way unequal."* That is, each number represents sides of a rectangle. Calculating the area of such a rectangle (4,800 × 2,700) gives **12,960,000**, the solution of the Nuptial Number.

Correspondences Between the "Sacred Tree Script" and the Constitution of the Universe

There are two levels to comprehend the significance of the Nuptial Number and its relationship to the "Sacred Tree Script": first, the importance of the final number 12,960,000; and, second, the meaning of the equations and numbers which were used to derive or calculate this number. Before discussing the number, 12,960,000, we would mention one point in relation to the number 216, which was calculated from Part A. Although Adam (Adam, 1985) suggested that the significance of this number was the relationship to the human period of gestation, it is also interesting that 216 is the exact number of syllables contained in the first sukta of Rig Veda (twenty-four syllables of the first richa plus the remaining 192 syllables – refer to Figure 1 for details). Perhaps Plato was referring to the structure of the Constitution of the Universe when he wrote this in his Republic (the first sukta of Rig Veda could be thought of as a gestation period in which the embryo of the Constitution of the Universe is fully developed during 216 syllables).

The chief significance of the number 12,960,000 is its reoccurrence in the records of other ancient civilizations. For example, the Babylonian "world year" or "world cycle" is 12,960,000 days or 36,000 years (assuming 360 days per year). Further, the key unit of the Babylonian sexagesimal system is 60 which when raised to the fourth power (the final step in the derivation of the Nuptial Number from Part B) gives 12,960,000. The Vedic Literature also makes reference to this number (Maharishi, 1967, Chapter 4, verse 1). According to Vedic chronology, there are four time periods or "yugas" which repeat themselves in a continuous cycle (chaturyugi): sat-yuga (1,728,000 years), treta-yuga (1,296,000 years), dvapara-yuga (864,000 years) and kali-yuga (432,000 years). As is evident, the 1,296,000 years for treta-yuga is similar to Plato's Nuptial Number of 12,960,000. Also, three chaturyugis (3 × 4,320,000) equals 12,960,000 years.

Another important aspect of the final Nuptial Number, 12,960,000, is its relationship to an octaval structure which is latent in Plato's "Great Year." According to Adam (Adam, 1985), the "Great Year" consists of 12,960,000 years and represents the time period required for all *eight* spheres to make one complete cycle and reach the same point from where they began. These *eight* spheres are as follows: the fixed stars, Saturn, Jupiter, Mars, Mercury, Venus, Sun and the Moon. It is worth speculating that these eight spheres may not represent the physical planets them-selves, but rather eight basic or fundamental aspects of natural law.

The next level of interpretation is to examine the equations and numbers that were used to derive the number 12,960,000. From the work of Efron (Efron, 1941), there are four significant aspects in the Nuptial Number derivation corresponding to the numerical structure of the "Sacred Tree Script." These four aspects provide "indirect evidence" that Plato may have formulated his description of the Nuptial Number according to the struc-ture of the "Sacred Tree Script."

First, it is evident from Parts A and B in the above discussion that the Pythagorean triangle formed a key element in the derivation of the Nuptial Number. It turns out that the numbers three, four and five which define the Pythagorean triangle precisely correspond to the numerical character-istics of the upper portion of the "Sacred Tree Script" (refer to Figure 34). Numbers three and four, representing two sides of the Pythagorean triangle, match the three twigs on the upper left side and the four twigs on the upper right side of the tree symbol. These two sets of twigs occupy the same space on the tree and below them is located the fifth twig on the right side. This fifth twig is large, thick and corresponds to the number five (equal to the hypotenuse of the Pythagorean triangle).

The second parallel with the "Sacred Tree Script" describes Plato's two "harmonies" used in the derivation of the Nuptial Number. As stated

Figure 34. Locating the Pythagorean Right-Angled Triangle in the Upper Portion of the "Sacred Tree Script"

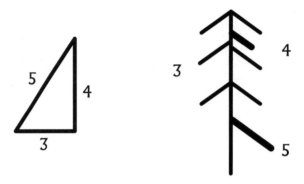

Adapted from Efron, A. (1941). *The Sacred Tree Script: The Esoteric Foundation of Plato's Wisdom.* New Haven, Connecticut: The Tuttle, Morehouse and Taylor Company, p. 137.

above, these two "harmonies" are as follows: 4800×2700 and 3600^2. If we remove the zeros from these three numbers (i.e., to get forty-eight, twenty-seven and thirty-six), it becomes possible to make an important link with the "Sacred Tree Script." The numbers forty-eight, twenty-seven and thirty-six, when expressed in the "Sacred Tree Script" notation, represent "harmonious" tree numbers (see Figure 35). "Harmonious" tree numbers are defined as those numbers which when written in the "Sacred Tree Script" notation have either an equal distribution of twigs on both sides or the presence of all the twigs on one side of the tree symbol. It is also interesting to note that the number thirty-six can also be derived directly from the octaval counting system (i.e., $1 + 2 + 3 + 4 + 5 + 6 + 7 + 8 = 36$).

The third significant correspondence between the Nuptial Number and the "Sacred Tree Script" is hidden in Plato's curious way of arriving at the number 4800. According to the excerpt in Part D (see above), one of the two equations used to calculate the number 4800 is as follows:

$$(7^2 \times 100) - (1 \times 100) = 4,900 - 100 = 4,800$$

In this equation, as well as elsewhere in his science of numbers, Plato acknowledges the importance of the number seven. Similarly, in the "Sacred Tree Script" the number seven, represented by the seventh twig on the right

Figure 35. Expressing the Two Harmonies of the Nuptial Number in the "Sacred Tree Script" Notation

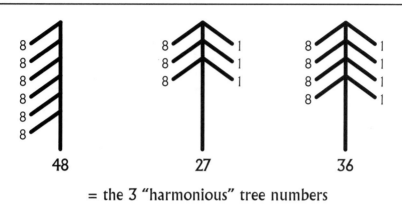

= the 3 "harmonious" tree numbers

Adapted from Efron, A. (1941). *The Sacred Tree Script: The Esoteric Foundation of Plato's Wisdom.* New Haven, Connecticut: The Tuttle, Morehouse and Taylor Company, p. 130.

side, delineates an important numeral. The importance of this seventh twig is evident from its length and size in relation to the other twigs.

The final comparison relates the "Sacred Tree Script" to the expression of the three counting systems displayed in the Nuptial Number. First, the use of the decimal system is obvious, in the final number 12,960,000, as well as in the number 3,600. Additionally, the numbers five, fifty and 100 are present in some of the equations used to calculate the Nuptial Number. The second counting system is the Babylonian sexagesimal system. In the final step in the derivation of the Nuptial Number, from Part B, the number sixty is raised to the fourth power. The number sixty represents the key unit of the Babylonian sexagesimal system. The last counting system is the octaval system. Evidence for employing this number system in the Nuptial Number was presented earlier. Both the Pythagorean right-angled triangle and the "harmonious" tree numbers thirty-six, twenty-seven and forty-eight represent special numbers in the octaval "Sacred Tree Script." These three counting systems, which are used to formulate the Nuptial Number, turn out to be the exact same three counting systems inherent within the "Sacred Tree Script." As discussed earlier, these three counting systems also correspond to the qualities of rishi, devata and chhandas from the Rig Veda.

In light of these intriguing correspondences, it seems likely that Plato knew about the octaval structure of the "Sacred Tree Script" when he formulated his Nuptial Number. Although the evidence presented by Efron (Efron, 1941) is indirect, it provides new insights into the obscure writings of Platonic wisdom. Furthermore, Efron's research carries additional importance considering that the numerical structure of the "Sacred Tree Script" is precisely the same as that of the Constitution of the Universe. The possibility that Plato incorporated the structure of the Constitution of the Universe into his science of numbers may explain the antiquity of Platonic wisdom and the depth and range of influence of his philosophy.

SECTION VII

ANCIENT MAYANS – TZOLKIN,
THE SACRED MAYAN CALENDAR

The word "Maya"[6] derives its origin from "Mayab" which describes the Yucatan Peninsula in Central America, a key area of the ancient Mayan civilization. Within this region, as well as scattered throughout the Mesoamerican jungles, are numerous ancient cities and temples intricately incised with elaborate hieroglyphics and astronomical data. These remnants signify that the Mayans possessed a seemingly high degree of knowledge in time-keeping, mathematics and astronomy. Even though the classical period of the Mayans was relatively short, 435 to 830 AD, they acquired an elaborate wisdom regarding the sacred order and celestial cycles of the cosmos. Much of this knowledge is embodied in one of the most outstanding achievements of this civilization, the Sacred Mayan calendar which played a significant role in everyday Mayan life as well as in cultural/religious activities (Arguelles, 1984). In this section we take a close look at the structure of the Mayan Calendar and present evidence, primarily derived from the work of Jose Arguelles (Arguelles, 1984; Arguelles, 1987), that the Constitution of the Universe can be located in this Calendar.

The Mayan Conception of Time and their Calendar Systems

Every society in ancient Mesoamerica possessed some type of calendar system that served as one of their most basic cultural items. The earliest evidence of calendar use appears to have originated in the Olmec Civilization which preceded the Classic Mayan Civilization by several

---

[6] It is of interest to note that words very similar to "Maya" are found in many other cultures (Arguelles, 1987, p. 17). For example, in India the Sanskrit word "Maya" is translated as "world of illusion" or "origin of the world." Also, the name of Buddha's mother was Maya. In Egyptian philosophy the word "Mayet" means "universal world order." Our month of May is derived from the Roman goddess, Maia or "the great one." In Greek Mythology the brightest star in the constellation Pleiades is called Maia.

centuries. However, it is the Mayans who have been widely accredited with having perfected the use of calendar systems. Even today there is a virtually unbroken sequence of calendar inscriptions dating from the early Mayans that is still used in some parts of Central America. This is fortunate considering the lapse of time and destruction caused by later Spanish invasions (Anton, 1978).

The Mayans developed several complex ways to measure time, each of them being intimately related to one another, which served both practical and cultural/religious purposes. In terms of time-keeping, the day or "kin" is considered the basic unit. Built upon the "kin" are various larger units including the following (Morley and Brainerd, 1983):

1 Kin	=	1 day		
1 Virnal	=	20 Kins	=	20 days
1 Tun	=	18 Virnals	=	360 days
1 Katun	=	20 Tuns	=	7,200 days
1 Baktun	=	20 Katuns	=	144,000 days
1 Great Cycle	=	13 Baktuns	=	1,872,000 days

The "Great Cycle" is equal to approximately 5,125 Earth years. The beginning of the present "Great Cycle" is estimated (the exact date is still disputed) between August 6 and August 13, 3113 BC, and will be completed on December 21, 2012 AD.

Knowledge of the Mayan calendar systems was considered the most important science and was undoubtedly held in high regard by the ruling class. It appears that there were at least four different calendar systems used by the ancient Mayans (De Paz and De Paz, 1993). In general, the numerical notation used in these systems is based on the vigesimal system (base twenty: 1, 20, 400, 8,000, 160,000, etc.) in which three symbols are used to signify numbers: the dot (representing one), the bar (representing five) and the shell (representing zero).

The Sacred Calendar or ritual almanac, based on a 260-day cycle, was the basic system used for determining the events of ceremonial life, prophecy and spirituality. What the Mayans actually called their Sacred Calendar is not known. Today archeologists commonly refer to it, as will be done here, as the Tzolkin. Tzolkin is a Yucatec word meaning "Book of Days" or "Count of Days." The second time-measuring device used by the

Mayans, the HAAB or tropical calendar consists of a 365-day cycle, eighteen months of twenty days each plus five additional days. This calendar recorded the cycles of the seasons. The third way of measuring time, the Calendaric Round, combined the Tzolkin and HAAB. That is, the 260 days of the Tzolkin joined with the 365 days of the HAAB to produce a total of 18,980 different days, or 52 years (365 days each year), in each calendar cycle. Interestingly, fifty-two years is the exact time period that it takes the constellation Pleiades to cross the zenith. The last time-device the Mayans employed has been called the Long Count which records time in a linear way, calculating the number of days since the Mayans began to count time. In this section we limit our discussion to the Tzolkin or Sacred Calendar because prophecy, religious acts and ceremonial life were governed based on this system. In addition, the Constitution of the Universe can be located within the Tzolkin.

The "Mystic Column" and "Loom of Maya" in the Tzolkin

The Tzolkin is a calendar system constructed from twenty sacred symbols interlocking with thirteen numbers to create a total of 260 different days (refer to Figure 36). The twenty different sacred symbols are as follows: Imix, Ik, Akbal, Kan, Chicchan, Cimi, Manik, Lamat, Muluc, Oc, Chuen, Eb, Ben, Ix, Men, Cib, Caban, Etznab, Cauac and Ahau. These twenty Mayan day-signs or ideoglyphs endlessly repeat themselves. Each of these sacred symbols is combined with a number, ranging from one to thirteen, to create the 260 days of the Tzolkin. Every day of this sacred calendar system had a unique meaning and represented an abstract principle or quality of Nature.

For purposes of clarity, the Tzolkin is presented in a grid lattice (Figures 37 and 38), after the style used by Arguelles (Arguelles, 1984; Arguelles, 1987). Although this grid appears to be a mere set of numbers, it embodies an almost magical order. For example, each number is always surrounded by the same set of numbers (e.g., the number 8 is always surrounded by the numbers 13, 7, 1, 2, 3, 9, 2 and 1). Down the middle of this calendar grid lattice, column seven, is the "Mystic Column" or "Axis of the Eternal Present." As we will describe shortly, on either side of this axis a symmetrical pattern is located. Between the number thirteen and the number one, on the "Mystic Column," lies the "great void" or the

Figure 36. A Mechanistic Representation of the Tzolkin

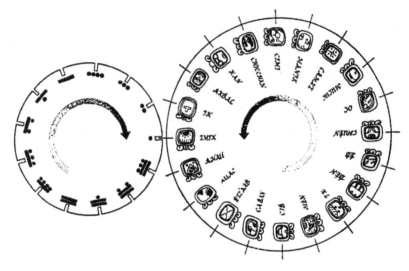

The Tzolkin is constructed of 20 symbols interlocking with 13 numbers:

Twenty Sacred Symbols					Numbers 1 through 13				
1. Imix	2. Ik	3. Akbal	4. Kan		1 •	2 ••	3 •••	4 ••••	
5. Chicchan	6. Cimi	7. Manik	8. Lamat	**×**	5 ——	6 —•—	7 —••—	8 —•••—	9 —••••—
9. Muluc	10. Oc	11. Chuen	12. Eb						
13. Ben	14. Ix	15. Men	16. Cib		10 ═══	11 ═•═	12 ═••═	13 ═•••═	
17. Caban	18. Etznab	19. Cauac	20. Ahau						

Equals a Total of 260 Combinations
(1 Imix, 2 Ik, ..., 13 Ben, 1 Ix, ..., 7 Ahau, 8 Imix, ...)

Adapted from Coe, M.D. (1971). *The Maya.* New York: Praeger Publishers.

"point of total balance" (Shearer, 1975). In this great void the Mayans conceived of time as an "endless path," with each time period corresponding to a "piece of that endless path" (De Paz and De Paz, 1993). The numbers thirteen and one mark the "end" joined to the "beginning." This notion of time matches the first unit of the Constitution of the Universe, represented in Rig Veda by "A-K." According to the Vedic Literature, "A"

Figure 37. Locating the Constitution of the Universe in the Tzolkin

20 Day Symbols:

"Mystic Column" / "Axis of Eternal Present"

Day Symbol												
Imix	1	8	2	9	3	10	4	11	5	12	6	13	7
Ik	2	9	3	10	4	11	5	12	6	13	7	1	8
Akbal	3	10	4	11	5	12	6	13	7	1	8	2	9
Kan	4	11	5	12	6	13	7	1	8	2	9	3	10
Chicchan	5	12	6	13	7	1	8	2	9	3	10	4	11
Cimi	6	13	7	1	8	2	9	3	10	4	11	5	12
Manik	7	1	8	2	9	3	10	4	11	5	12	6	13
Lamat	8	2	9	3	10	4	11	5	12	6	13	7	1
Muluc	9	3	10	4	11	5	12	6	13	7	1	8	2
Oc	10	4	11	5	12	6	13	7	1	8	2	9	3
Chuen	11	5	12	6	13	7	1	8	2	9	3	10	4
Eb	12	6	13	7	1	8	2	9	3	10	4	11	5
Ben	13	7	1	8	2	9	3	10	4	11	5	12	6
Ix	1	8	2	9	3	10	4	11	5	12	6	13	7
Men	2	9	3	10	4	11	5	12	6	13	7	1	8
Cib	3	10	4	11	5	12	6	13	7	1	8	2	9
Caban	4	11	5	12	6	13	7	1	8	2	9	3	10
Etznab	5	12	6	13	7	1	8	2	9	3	10	4	11
Cauac	6	13	7	1	8	2	9	3	10	4	11	5	12
Ahau	7	1	8	2	9	3	10	4	11	5	12	6	13

■ = "Loom of Maya"
(refer to Figure 38 for details)

Reproduced with modifications from Arguelles (1984 and 1987).

Figure 38. The "Loom of Maya" of the Tzolkin

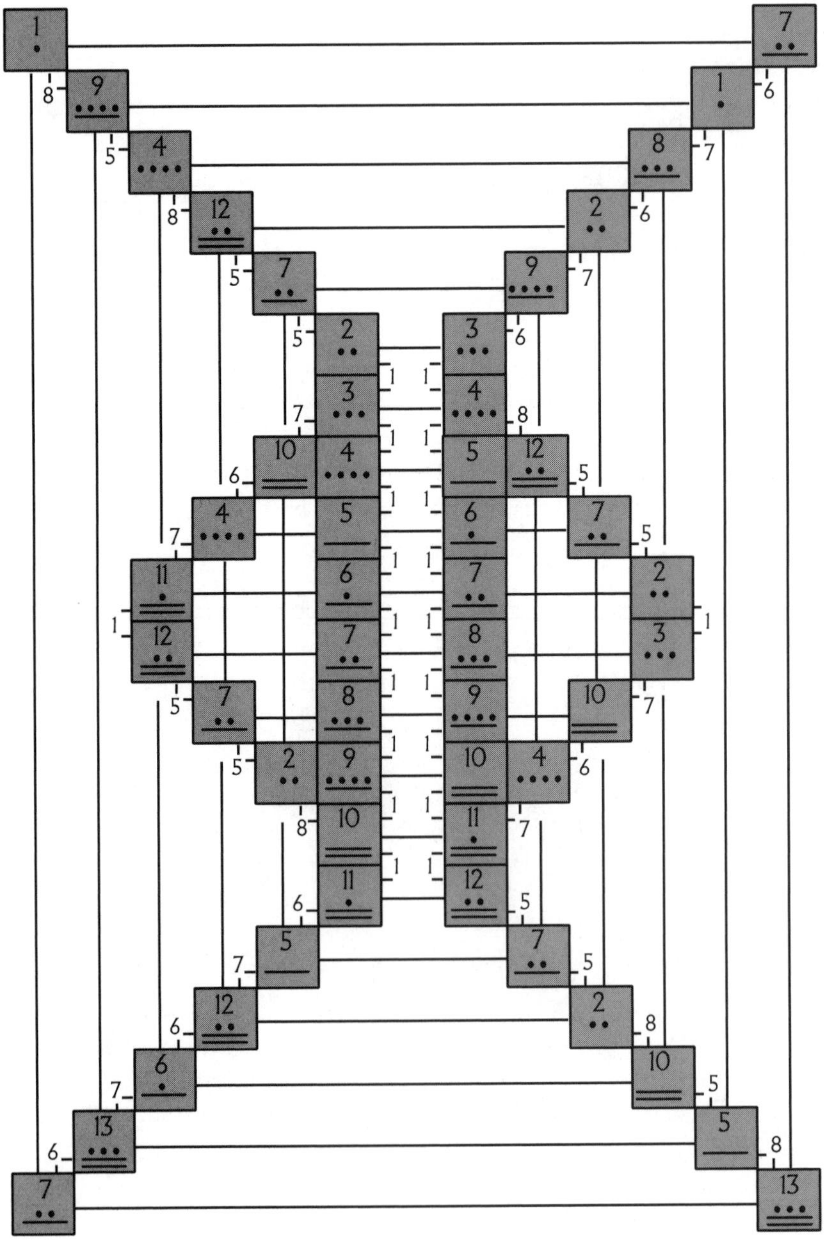

Reproduced with modifications from Arguelles (1984 and 1987).

represents the Absolute or Ultimate Reality, which collapses to the point value of "K." Contained within "A-K" is the seed form of all knowledge.

From this "great void" or "point of total balance" the Tzolkin was presumably constructed, a calendar system which was to reflect the sacred order of the cosmos. One of the key images embedded in this calendar matrix is a magic fifty-two-unit design that first appeared in the work of Tony Shearer (Shearer, 1975, p. 82). Later, Jose Arguelles (Arguelles, 1984 and 1987) elaborated upon this image and coined the terms, the "Loom of Maya" or the "Binary Triplet Configuration," to describe its geometric symmetry and basic importance to the Tzolkin. This fifty-two-unit pattern (refer to Figures 37 and 38) provides the unifying structure for weaving together the thirteen numbers and twenty symbols to produce the complete 260-unit Sacred Calendar matrix. As mentioned earlier, fifty-two is also significant because it defines the total number of years in each cycle of the Calendaric Round that displays the combination of the Tzolkin and the HAAB calendar systems.

The entire "Loom of Maya," consisting of a total of fifty-two units, conveys a bilateral or twofold symmetry defined by twenty-six units of the image lying on either side of the "Mystic Column." The "Mystic Column" provides an axis of symmetry for either side of the calendar. As well written by Arguelles, "The invisible seventh is the mystic column. Unmirrored, it mirrors all." (Arguelles, 1987, p. 89). The binary nature of this image is parallel to that of the I Ching (refer to Chapter 8, Section IV) binary system of the broken (yin, negative) and unbroken (yang, positive) lines forming the eight trigrams and the sixty-four hexagrams.

The left side of the "Loom of Maya" is said to represent a negative and downward energy flow, while the right side represents positive and upward energy flow (Arguelles 1984 and 1987). This is analogous to DNA where both strands of the double helix run in opposite directions and one strand is the template for the other strand. In addition, the image of the "Loom of Maya" matches the two strands of DNA winding and crossing around one another. This pattern of "crossover" is also evident in the interval numbers in the "Loom of Maya" (Figure 38). Each interval number is calculated by determining the difference between any two adjacent numbers on the "Loom of Maya" configuration. For example, the difference between one and nine, the first two units in the upper corner of the "Loom of Maya," is eight. By examining these interval numbers in Figure 38, it is evident that there are two patterns of numbers: "8, 5, 8, 5 and 5" and "6, 7, 6, 7 and 6" which characterize the two strands of the "Loom of Maya," similar to the two strands of DNA.

Upon further examination of the "Loom of Maya," we find that it can be broken down into thirteen sets of four units each, starting with each corner and moving inward one square at a time for each set of four (Figure 38). For example, the first set equals: 1, 7, 13 and 7 (arithmetic total of 28); the second set equals: 9, 13, 5 and 1 (arithmetic total of 28); and the last set (in the center): 6, 7, 8 and 7 (arithmetic total of 28 again). As is readily apparent, all of these thirteen sets of four numbers add up to twenty-eight, the number of days in a lunar month. Multiplying twenty-eight by the "magic" number of thirteen (used in the construction of the Tzolkin) gives 364, approximately the number of days in a lunar year.

After thirty-three years of research on the ancient Mayans, Jose Arguelles discovered that it was this "Loom of Maya" which links the Tzolkin with the sixty-four hexagrams of the I Ching, with the sixty-four codons of DNA and with the 8 × 8 Benjamin Franklin Magic Squares (Arguelles, 1984 and 1987). Because we have already linked the I Ching (Chapter 8, Section IV), the genetic code of DNA (Chapter 4, Section I) and Magic Squares (Chapter 7, Section II) with the Constitution of the Universe, it was an obvious next stage to use Arguelles's work to connect the Tzolkin with the structure of the Rig Veda.

Arguelles located the number sixty-four, corresponding to the sixty-four hexagrams of the I Ching, in the center of the Tzolkin (Figure 37), accommodating the "Loom of Maya" crossover pattern. This sixty-four-unit matrix is referred to as the "crossover-polarity zone," with thirty-two units existing on either side of the "Mystic Column." This pattern is similar to the double helix crossover pattern characteristic of DNA. According to Arguelles, "The 64-unit "keyboard" is the genetic matrix of transformation which unified the entire 260-unit Tzolkin" (Arguelles, 1987, p. 161). All of the other units in the Tzolkin apparently represent elaborations on the central sixty-four units. It is worth noting that in any 8 × 8 (a sixty-four matrix) Magic Square (for more details refer to Chapter 7, Section II), the arithmetic sum of all vertical, horizontal and diagonal rows, each add up to 260 (the number of days in the Tzolkin).

The Tzolkin Contains the Same Structure as the Constitution of the Universe

Based on this initial work of Arguelles, we decided to further the parallels with the sixty-four-unit central matrix of the Tzolkin to include the Constitution of the Universe. As discussed earlier, the first unit of the Constitution of the Universe, "A-K," is located in the center of the Tzolkin between the numbers thirteen and one, representing the "point of total balance" or the "great void." The next level in the Constitution of the Universe is the unfoldment of the first syllable "A-K" into the eight sylla-bles of the first pada (Figure 1). These basic eight syllables of Rig Veda correspond to the eight units on the "Mystic Column" or "Axis of the Eternal Present", reflecting the crossover symmetry of the "Loom of Maya" (refer to Figure 37). These eight units of the "Mystic Column" act as an axis of symmetry for the thirty-two units located on either side of this column which together comprise the sixty-four-units of the central matrix. This sixty-four-unit central matrix corresponds precisely with the final level of the Constitution of the Universe, the sixty-four units unfolding from the eight syllables of the first pada (see Figure 1).

According to Arguelles (Arguelles, 1984) the ancient Mayans believed that the principle of heaven, man and earth permeates every phenomenon and object. These three fields do not exist apart from one another, but form an ever-present axle of support. Heaven is above, earth is below and man forms the link between them. This understanding is similar to the three values of rishi (knower), devata (process of knowing) and chhandas (known) found in the Constitution of the Universe. The first principle described by the Mayans was heaven. This matches the abstract and spiri-tual nature of rishi. The second principle, "man," was believed to link heaven and earth and corresponds to the nature of devata. Devata connects or links rishi to chhandas. The last principle, earth, corresponds to the material and concrete nature of chhandas. Together these three Mayan principles provide a complete match between the Tzolkin and the Constitution of the Universe. Each of the eight units on the "Mystic Column" and the sixty-four units that lie on either side of the "Mystic Column" may be viewed with respect to heaven, earth and man, thus comprising a total of twenty-four units and 192 units respectively. In the Constitution of the Universe this corresponds to the eight prakritis and their elaboration into sixty-four aspects viewed with respect to rishi, devata and chhandas (comprising twenty-four and 192 units respectively).

In light of the interesting parallels presented in this section, it seems likely that the Tzolkin was more than just a time-keeping device or "agricultural calendar." Yet, even if we examine the Tzolkin solely from the angle of astronomical properties, it is impressive. For example, the Mayans were able to calculate lunar cycles, predict eclipses, compute the Earth's revolution around the sun (in days) to within a thousandth of a decimal point, and record the movements of the inner planets. In addition to these calendar achievements, the probable links between the Tzolkin and the Constitution of the Universe suggest that there is an even deeper significance and meaning to the Tzolkin which could help explain why the ancient Mayans had such an obsession with time and invested so much in their calendar systems.

PART B

THE GENESIS OF THE MATERIAL UNIVERSE

The chapters included in this second part of the book present an understandable and elegant restoration of the ancient wisdom regarding the genesis of the material universe. Our analysis of this topic is primarily derived from the knowledge contained in the Vedic literature, which presents a vivid account of the exact mechanics responsible for governing all material objects. For the most part, the clear systematic understanding of this process has remained obscure and covert. Although many of the individual concepts we will be discussing are widely known, no one to our knowledge has synthesized these partial values to create a unified perspective.

The earlier part of this book was devoted to understanding the fundamental patterns and information in the Constitution of the Universe. Both the material and subjective aspects of our universe are created from this universal constitution. The purpose of this part of the book is to derive, from the Constitution of the Universe, the exact mechanics responsible for the creation of the material universe. In the first chapter we describe the primal mechanics for the creation of all material objects as understood by the ancient Vedic civilization. The eight subsequent chapters present definitive evidence that the exact numerical structure of this mechanics is found in the fundamental concepts of the academic disciplines and that it was understood by other ancient civilizations.

In brief, the primal mechanics responsible for the genesis of the material universe consist of a two stage process. The first stage in the creation of material objects is the emergence of five primary agents or elements from a unified source. In the second stage, these five fundamental elements pair in a well-defined pattern to create three operating principles. Collectively these five and three entities interact in numerous ways to produce all material diversity. Although this may at first sight seem too trivial, what we need to realize is that universal truth is, in fact, very simple. When we truly understand what the Ancients were talking about, it becomes clear that the mechanics governing our universe are not complex, but rather quite obvious and beautifully simple. We have found that the more we think about the research presented in this book, the more we realize that this knowledge precisely fits together the pieces of the cosmic puzzle that have remained vague and concealed for so long.

The topics comprising the chapters in this part of the book were selected because they provide clear illustrations of the exact mechanics of the genesis of the material universe. We realize that there are also many other examples, especially from other systems of knowledge and ancient civilizations describing the same basic mechanics. Unfortunately, however, we were unable to obtain detailed information from these sources, and in most situations the original teachings appear to have been distorted. For instance, the idea that the material universe is founded upon five primal entities was known throughout the ancient world, but over time this wisdom was lost and attempts to revive it have been incomplete (e.g., it is often thought that the universe was constructed on four primal elements instead of five). Based on our research and growing confidence, we feel that the information presented in this book provides a central reference which can be used to restore the teachings from the ancient world. Such in-depth research, however, is beyond the scope of this book. Instead our aim is simply to provide conclusive evidence for a unified understanding of the exact mechanics for the creation of the material universe.

In studying the genesis of the material universe, we suggest that the reader begin with the first chapter. Because the information is both simple and clear, it should be relatively easy to grasp the central concepts that will enable the reader to understand the significance and implications of all the following chapters. It is not necessary to read the subsequent chapters in sequence. In fact, we suggest reading first the chapters pertaining to familiar disciplines or areas of interest. Perhaps the greatest asset to understanding this material is to study the diagrams which provide a summary of everything we need to know.

CHAPTER 9:

ANCIENT VEDIC CIVILIZATION: THE FIVE TANMANTRAS MERGING TO FORM THE THREE DOSHAS

The ancient Vedic civilization contains one of the oldest, largest and unaltered collection of literary works, bigger than the records of all other ancient cultures combined. For this reason, the ancient texts of this civilization have become a valuable source for research and understanding the laws of nature governing our universe. The earlier part of this book used the wisdom contained in the Vedic literature to describe the Constitution of the Universe, a fundamental template from which the material and subjective aspects of our universe are created. In this chapter we build upon this knowledge and derive the primal mechanics responsible for the genesis of all material diversity from the Constitution of the Universe. Comprehending the information in this chapter provides the foundation for everything that follows.

Consciousness has Three Aspects: Rishi, Devata and Chhandas

In Figure 39 we have presented a summary of the essential mechanics responsible for the genesis of the material universe as understood by the ancient Vedic civilization. According to Rig Veda and other aspects of the Vedic literature all material diversity has its origin in an unmanifest field of universal intelligence or consciousness. This unified level of consciousness was believed to be fundamental to nature and was considered to be the lively origin and basis of everything in creation. Consciousness is a subjective level of intelligence that is devoid of the common, everyday experience of individualization and concrete experiences. Through direct cognition and exploration of this universal field of consciousness, the

Figure 39. Genesis of the Material Universe According to the Vedic Literature: The Three Doshas Arising from the Five Tanmantras

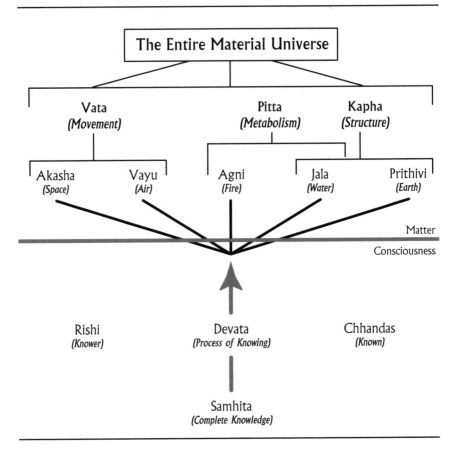

ancient Vedic seers were able to comprehend the fundamental mechanics upon which our universe is governed. The wisdom from these ancient cognitions was written down and passed on generation after generation to create what we know as the Vedic literature.

Due to its intrinsic characteristics, this universal field of consciousness has a threefold structure. According to Maharishi, this threefold structure of total knowledge or consciousness is derived from the very nature of consciousness itself (Maharishi, 1986; Maharishi, 1994). What we need to understand is that consciousness, simply by virtue of its own existence, is conscious or aware. The question then arises, "What is consciousness aware of?" If consciousness is the only one ultimate reality of all that there is in the universe, then consciousness can only be conscious of itself. This leads us to the intellectual necessity of a process through which consciousness can know itself and again this third concept can only be consciousness. In this way, we have derived the threefold structure of knowledge or consciousness. These three aspects of consciousness are referred to in the Vedic literature as rishi (knower), devata (process of knowing) and chhandas (known). Another way to describe this structure of consciousness is to say, consciousness (the knower) is aware of consciousness (the known) through the agency of consciousness (the process of knowing).

Each of the three concepts of consciousness has its own unique characteristics. The qualities often associated with the rishi (knower) aspect of consciousness include abstract, inherent nature, subjective and intelligence. Other meanings of the word rishi include "observer," "seer" or "silent witness." The second aspect of consciousness, devata (process of knowing), is often denoted by qualities such as dynamism, activity, transformation, change, interaction and movement. The last aspect of consciousness, chhandas (known), embodies characteristics such as material, concrete, physical, structural and worldly. In the Vedic literature these three aspects of knowledge collectively are known as the samhita (pronounced sang-heeta) value or the complete structure of knowledge. Samhita, therefore, is the togetherness of rishi, devata and chhandas.

It is important to realize that the three aspects of knowledge are inseparable from each other and should not be regarded as isolated entities. In deriving the threefold nature of consciousness, we are not creating anything new, but rather describing consciousness from three different viewpoints. What this means is that the three different interpretations or ways of describing consciousness are merely an intellectual conception. At the bottom of Figure 39 the samhita value (complete knowledge) is located along with the three aspects of consciousness, rishi (knower), devata (process of knowing) and chhandas (known). From this level or structure of consciousness the entire universe is created in such a way that this threefold structure of knowledge is expressed in all entities throughout the diversified layers of creation.

The genesis of the universe is divided into two aspects: the creation of the material universe and the creation of the subjective universe. Both processes are derived from the Constitution of the Universe. As the reader may remember, the first stage of unfoldment in the Constitution of the

Universe is the creation of the eight primal tendencies of nature (the eight prakritis). These eight prakritis are divided into two groups: the first three are responsible for the subjective aspects of life, and the last five are responsible for the objective or material aspects of life. The subjective universe emerges from the first three prakritis and the material universe emerges from the last five prakritis. For the purpose of this book, we have chosen to present the exact mechanics for the creation of the material universe and not the subjective universe because most of the information available in our society today is concerned with the material universe. Such a detailed body of knowledge and information is necessary to validate the existence of the primal mechanics of the universe throughout the academic disciplines. Unfortunately, at this time there is little knowledge available regarding the detailed nature of what constitutes the subjective universe.

The Five Tanmantras are the Primordial Governing Agents

The very first stage in the creation of the material universe is the formation of five fundamental entities from the universal field of consciousness and intelligence (refer to Figure 39). In the Vedic literature these primal entities are known as the five tanmantras or five subtle elements: Akasha (space or ether), Vayu (air), Agni (fire), Jala (water) and Prithivi (earth). The English translations given in parentheses do not connote the full values of these Sanskrit terms, but rather serve to give only an idea or flavor of what that tanmantra really represents. For instance, if we were to ask a Vedic sage what each of the five tanmantras describe, we would not expect him to point to a burning log, the sky, the wind or a creek. Instead, the Vedic sage would likely say that the tanmantras are primordial governing agents or forces responsible for creating the material universe. The qualities associated with each tanmantra can be derived from the characteristics embodied by the English translations given above. The following is a list of some of the inherent qualities for each of the five tanmantras:

- Akasha, the first tanmantra, translates as the space or ether element. Qualities exhibited by this tanmantra are emptiness, clearness, vastness, voluminous, abstract, pervasive and subtle. It is a background or medium for existence, change and dissolution.

- Vayu, the second tanmantra, translates as the air element. Characteristics often associated with this tanmantra are movement, transportation, change, wind, light weight, quick, lucid and influential.

- Agni, the third tanmantra, translates as the fire element. Attributes associated with the Agni tanmantra are things such as heat, digestion, transformation, metabolism, brightness, lightness, destruction and sharpness.

- Jala, the fourth tanmantra, is translated as the water element. Qualities often associated with the Jala tanmantra are the force of cohesion (e.g., a water droplet), the power of attraction, fluidity, moisture, liquidity and flowing.

- Prithivi, the fifth tanmantra, translates as the earth element. Characteristics embodied by the Prithivi tanmantra are structure, solidity, heaviness, sturdiness, denseness and physical matter.

The qualities and attributes associated with the tanmantras are meant only to give the reader an idea or feeling for the nature of these primal entities. In reality they are fundamental governing agents with primordial characteristics which dictate how and what happens in the material universe. These five tanmantras are not only described in the opening of the Rig Veda (as part of the Constitution of the Universe), but are also widely recognized both in the popular Ayurvedic literature as well as the classical texts of Ayurveda. (Ayurveda is the medical system of the ancient Vedic civilization and its texts make up an important part of the Vedic literature.) In Ayurveda the individual is said to be composed of the five tanmantras and that imbalances in any one of the tanmantras give rise to disease. According to Ayurveda the five tanmantras give rise to the five senses which allow us to perceive the external world in five distinct ways to construct our view of reality. As will become evident in following chapters, the five tanmantras are not only mirrored throughout creation, but they also provide the basis for dividing, grouping and classifying the entire universe.

The Five Tanmantras Combine to Form the Three Doshas

In the second stage in the creation of the material universe the five tanmantras are paired in a well-defined pattern to create three new operating principles (refer to Figure 39). In the Vedic literature these three operating factors are called the three doshas: vata (movement), pitta (metabolism) and kapha (structure). The three doshas give form to everything in the material universe, and through their constant shifting create new forms and images. Below we have described how the tanmantras combine to create the doshas, as well as the qualities associated with each dosha:

- Vata (movement) dosha is created from Akasha (space) and Vayu (air). Because vata dosha is composed of air and space, it has the qualities of change, dryness, cold, rough, minute, quick, unpredictable, light weight and clear. The major functions of vata dosha are movement, transportation and communication.

- Pitta (metabolism) dosha is created from Agni (fire) and an aspect of Jala (water). Since pitta dosha is derived from fire and water, it is endowed with characteristics such as intense, sharp, precise, hot, moist, fluid, scorching and flowing. The major functions of pitta dosha are metabolism, digestion and transformation.

- Kapha (structure) dosha is created from Jala (water) and Prithivi (earth). Because kapha dosha is derived of water and earth, it has attributes such as slow, heavy, relaxed, sturdy, steady, soft, smooth, dense, unctuous, sticky and solid. The major functions of kapha dosha are structure and cohesion.

The exact nature of the pairing of the five tanmantras to create the three doshas is described in the classical texts of Ayurveda (Sharma and Dash, 1988) and has been clearly defined in the recent scientific literature (Lad, 1984; Frawley and Lad, 1986; Hagelin, 1987; Hagelin, 1989; Chopra, 1990). One important issue, however, that needs clarification is a confusion in the more popular Ayurvedic literature regarding the generation of pitta dosha. It is thought by some individuals that pitta dosha is created from only one tanmantra, Agni (fire). Instead, the correct understanding is that pitta dosha is generated by the pairing of two tanmantras: Agni (fire) tanmantra and an aspect of Jala (water) tanmantra. Genesis of pitta dosha from the pairing of these two tanmantras is supported not only by the authoritative texts of Ayurveda, but also by the confirmatory evidence presented in the following chapters. Thus, changing this pattern of tanmantra pairing to form the three doshas would contradict the arrangements found to be naturally occurring in the laws of the material universe that are established by the academic fields.

In the ancient medical science of Ayurveda the three doshas are said to shape our personal reality, both physically and mentally. They represent the three basic types of human constitutions and regulate thousands of separate functions in our physiology and psychology. Together these three doshas are central to understanding the causes of disease and the remedies for health care. Ayurveda also recognized that these three doshas regulate the qualities associated with the times of day, the nature of seasons, and all objects in the material universe.

Collectively, the five tanmantras and three doshas regulate and construct the material universe without being quite physical themselves. They are constantly shifting to create new images and are increased and decreased in different proportions to create the mass diversity surrounding us. In this way, the entire universe of material objects can be grouped, classified and divided according to the nature of these primal entities.

The following chapters in this second half of the book provide eight clear illustrations that this pattern of the primal mechanics responsible for the genesis of the material universe is found in the fundamental aspects of the academic disciplines. To our knowledge, there has been only one piece of research published in this area relating to the fundamental classifications of matter and energy as understood by modern physics (refer to Chapter 10 for a detailed review). However, as a result of our research during the past two years, we have discovered seven other clear examples of this fundamental structure that gives substantial credence for the wisdom of the ancient Vedic culture.

CHAPTER 10:

PHYSICS: THE FIVE SPIN-TYPES MERGING TO FORM THE THREE SUPERFIELDS

During the last several decades, physicists have formulated mathematical theories that are beginning to fulfill Einstein's dream of finding a completely unified understanding of the ultimate origin and source of all entities in nature. These mathematical theories attempt to explain or accommodate the great diversity of particles and forces observed in our universe as arising from a single unified source. Some of the most successful and promising frameworks to emerge from these efforts are known as the superstring theories. Earlier in this book we presented a conceptual understanding of superstring theories and demonstrated that the mathematical structure of such formulations was identical to the numerical unfoldment of the Constitution of the Universe. In this chapter we build upon this knowledge to reveal yet another correspondence between ancient Vedic wisdom and modern physics. This new correspondence, originally published by Hagelin (Hagelin, 1987; Hagelin, 1989), will cover two primary topics. First, we will summarize the fundamental classification of particles and forces described by the currently accepted unified field theories of quantum physics. Second, we will demonstrate that the derivation of these primal classifications of physical matter exactly matches the structural mechanics responsible for the genesis of the material universe, as recorded in the ancient Vedic Literature. To our knowledge, this correspondence is the only piece of published research linking the ancient understanding of the genesis of the material universe to the fundamental aspects of the academic disciplines.

The Unified Field (Samhita) has Three Characteristics

Figure 40 presents an overview of the correspondences between the ancient Vedic understanding of the formation of the material universe and

Figure 40. Genesis of the Physical Universe According to Physics: The Three Superfields Arising from the Five Spin-Types

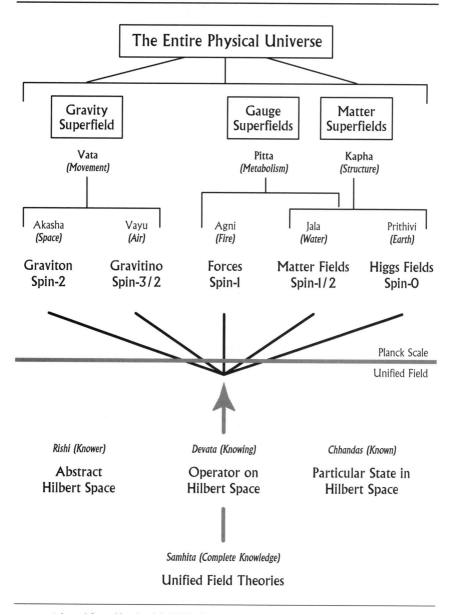

Adapted from Hagelin, J.S. (1989). *Restructuring Physics from its Foundation in Light of Maharishi's Vedic Science.* Modern Science and Vedic Science, 3 (1), pp. 2-72.

the accepted classification of particles and forces according to modern physics, which will be discussed in this chapter. We begin with a consideration of the unified field itself at the bottom of this figure. The unified field, the current term used in quantum physics to represent the totality of natural law, is the stable foundation on which the entire material universe is constructed. Mathematical theories describing the unified field provide an objective understanding of elementary particles and forces as stable modes or vibrational states of the field. However, only with the advent of superstring theories has modern physics been able to construct completely unified field theories. Now all known particles and forces of our universe can explicitly be identified with vibrational modes or expressions of a single unified field. Each force and particle can clearly be linked with a particular vibrational state of the field. In relation to the Vedic understanding of the genesis of the material universe, the unified field theories of modern physics correspond to the samhita value (total knowledge) because everything in physics is ultimately based on the concept of a unified field that is the origin and source of everything.

According to quantum field theory, the dynamics of the unified field can be described in three general ways satisfactorily corresponding to the flavors of rishi (knower), devata (process of knowing) and chhandas (known). The first way to describe the unified field is as an abstract field known as Hilbert Space. Hilbert Space, an infinite dimensional space in which all creation takes place, may be viewed as a field where all quantum mechanical functions live and exist. As an abstract vector space of all possibilities it can be expanded and interpreted in any other state. These descriptions of Hilbert Space correspond to the nature of rishi which is associated with knowledge and abstract information. The second way to view the unified field, called the Operators on Hilbert Space (e.g., energy, angular momentum, etc.), are dynamic generators of transformation in Hilbert Space. They actively transform one quantum-mechanical state into another and generate all change governing our universe. Without these Operators, and the transformation they generate, Hilbert Space would be completely inert. For these reasons, it makes sense to match the Operators in Hilbert Space with the dynamic and active nature of devata. The last way to view the unified field is to examine a particular state or individual point in Hilbert Space. A particular state of Hilbert Space represents an isolated possibility within the quantum mechanical field of all possibilities. Since each state corresponds to an actual mode of the physical system this matches the material nature of chhandas (Hagelin, 1987 and Hagelin,

1989). Together these three aspects of the unified field provide a profound and simple view of the quantum levels of life residing below the molecular, atomic and subatomic levels of creation.

From this unified field of perfect symmetry and complete unity the entire physical universe emerges through a process known as symmetry breaking. The principle of spontaneously broken symmetry locates deeply hidden symmetries of nature at fundamental space-time scales. During recent years physicists have successfully applied the principle of symmetry breaking to design unified field theories that combine previously unrelated concepts such as forces and particles into different modes of an underlying field. In this way, it has become possible to formulate theories to explain the rich emergence of diverse particles and forces from an initially unified state. At more fundamental levels we find increasing unity between the various particles and forces, whereas, at ordinary macroscopic scales we observe the superficial and apparent motion of diverse particles and forces.

Nature Limits the Number of Forces and Particles to Five Spin-Types

In unified field theory all elementary particles and forces belong to one of several fundamental categories distinguished by their quantum-mechanical spin. The simplest way to understand the spin of a particle is from the classical or macroscopic viewpoint. From this perspective one can imagine that particles are always found physically spinning and therefore possessing angular momentum. If these particles were to exist in the macroscopic world, they could possess any degree of speed and could, therefore, also possess arbitrary angular momentum. However, since elementary particles and forces reside at a quantum level, the concept of spin entails certain limitations with no classical analog. For instance, at the quantum level energy is found to be "quantized," meaning that it can be gained or lost only in small discrete "packages." The smallest unit of spin or discrete "package" is referred to as Planck's constant. Essentially this means that the magnitude of angular momentum of a spinning particle is quantized according to half-integer multiples of Planck's constant (e.g., 0, 1/2, 1, 3/2, 2, etc.). In other words, particles can spin only at units of Planck's constant, but not between these units. Because of the extreme smallness of Planck's constant, the quantum mechanical effects of spin are imperceptible at ordinary macroscopic scales.

To be consistent with the mathematical formulas of unified field theories, the number of forces and particles in nature is limited to five basic spin-types. Collectively, these five spin-types comprise the most fundamental classification of particles in physics, from which the entire physical universe is created. The unique spin of each particle determines, to a large degree, its statistical properties as well as many other physical character-istics. The five spin-types are known as follows: spin-0, spin-1/2, spin-1, spin-3/2 and spin-2. Spin-types greater than 2 do not lead to quantum mechanically consistent field theories. All spin-types arise from, and can be identified with, specific vibrational states or modes of an underlying string field, according to superstring theory. However, because of the tech-nical nature of this topic, we refer the interested reader to the work of Hagelin for a good discussion (Hagelin, 1987; Hagelin, 1989).

The Five Spin-Types are Grouped into Bosons and Fermions

The five spin-types or five fundamental classifications of all particles and forces can be further grouped into two general categories called bosons and fermions. Although bosons are traditionally associated with "forces," and fermions with "particles," unified field theories have established that both bosons and fermions can act as either particles or forces. Bosons include elementary particles characterized by integer spin-types (i.e., spin-0, spin-1 and spin-2). In general, bosons tend to occupy the same quantum state and thus usually exhibit coherent collective behavior (e.g., super conductivity, superfluidity and laser light). Examples of bosons include the four fundamental forces of nature: the massless photon responsible for electromagnetism; the $W^\pm$ and Z^0 bosons responsible for the weak force; the eight massless "gluons" of quantum chromodynamics responsible for the strong force; and, the massless graviton responsible for the force of gravity.

In comparison, fermions include elementary particles defined by half-integer spin-types (i.e., spin-1/2 and spin-3/2). As opposed to bosons, fermions are not allowed to occupy the same quantum state, according to Pauli exclusion principle, and thus cannot exhibit collective coherence. For instance, the fermionic property of electrons is responsible for the complex orbital structure of atoms giving rise to the diversity of chemical elements in the periodic table. Other examples of fermions include the proton and neutron which are the fundamental constituents of the atomic nucleus.

Because of the highly dissimilar properties of bosons and fermions, it has been difficult to unify these particles in order to design completely unified field theories. It was not until the advent of a mathematical symmetry known as "supersymmetry" that an ingenious framework for the unification of bosons and fermions was discovered. In its simplest form, supersymmetry unites particles of adjacent spin-types (i.e., spin-0 bosons with spin-1/2 fermions, spin-1/2 fermions with spin-1 bosons, etc.) into what are known as superfields or superstring fields. The unification of bosons and fermions through supersymmetry requires that both particles or forces have identical characteristics such as electric charge and mass. However, among the known elementary particles and forces there are no such pairs with identical physical characteristics. This means that the implementation of supersymmetry requires doubling the number of known particles in nature. In other words, we need to add a supersymmetric particle for each of the known particles and forces. Although there are no known examples of, or experimental evidence for, the existence of supersymmetric partners, there are a number of theoretical justifications for, and advantages to, using supersymmetry (refer to Hagelin, 1987 and Hagelin, 1989 for discussion). The primary value of supersymmetry is that it will completely unify opposite values, bosons and fermions, which provides great hope for constructing a completely unified field theory.

The Five Spin-Types Match the Five Tanmantras

Based on this background understanding, the following briefly describes the five fundamental quantum mechanical spin-types and their importance:

- The first category consists of spin-2 type particles known as gravitons. Gravitons are elementary particles or "quanta" that are responsible for the force of gravity and the space-time curvature. Although there is no experimental evidence for the existence of such particles, they are thought to be massless.

- The second category called the gravitinos includes all spin-3/2 elementary particles. Gravitinos are the supersymmetric partners of the spin-2 gravitons and are responsible for upholding local supersymmetry.

- The third category consists of spin-1 type particles responsible for the force fields. This group of elementary particles includes the electromagnetic force (mediated by the photon), weak force (mediated by the $W^{\pm}$ and Z^0 bosons), strong force (mediated by the eight gluons), and other superheavy grand unified particles. Collectively, all particles with spin-1 are commonly known as gauge bosons.

- The fourth category includes all the spin-1/2 elementary particles which are classified according to two groups. First, there is a group of particles known as the spin-1/2 gauginos that are the supersymmetric partners of the spin-1 force fields. Second, there is a collection of particles called the spin-1/2 matter fields (quarks, leptons and higginos) that are the supersymmetric partners of spin-0 matter fields.

- The fifth category consists of the spin-0 particles responsible for matter or scalar fields. Elementary particles with spin-0 have no intrinsic angular momentum. Included in this group are the Higgs bosons, sleptons and squarks which are the super-symmetric partners for the spin-1/2 higginos, leptons and quarks, respectively. One of the roles of the Higgs bosons is to give mass to all other particles when the fundamental symmetries of nature are broken (e.g., moments after the postulated Big Bang). Spin-0 particles are typically the most inert with respect to space-time transformation properties.

At this stage, we can draw our first major correspondence between the fundamental categories of matter and energy described by quantum physics and the exact mechanics responsible for the genesis of the material universe as understood by the ancient Vedic civilization. According to the Vedic Literature, there is a universal field of consciousness or intelligence that lies at the basis of creation. From this field of consciousness the first stage in the creation of the material universe is the emergence of five primal entities called the five tanmantras which interact to create the entire material universe. Similarly, in physics the unified field of all the laws of nature is considered to be the foundation of the entire physical universe. From the framework provided by unified field theories, such as superstring theory, all elementary particles and forces can be clearly categorized into five fundamental categories distinguished by their spin-type. In Figure 40 we have presented these five spin-types and their corresponding tanmantras as described by Hagelin (Hagelin, 1987; Hagelin, 1989). The reason the spin-types were matched with tanmantras in this order will become evident shortly.

Formation of the Three Superfields Corresponds to the Creation of the Three Doshas

The second stage in the creation of the material universe, according to the Vedic Literature, is the pairing of the five tanmantras in a well-defined pattern to create the three doshas. The pairing is as follows: akasha (space) and vayu (air) tanmantras merge to form vata (movement) dosha; agni (fire) and jala (water) tanmantras merge to create pitta (metabolism) dosha; and jala (water) and prithivi (earth) tanmantras merge to produce kapha (structure) dosha. Likewise, in the context of supersymmetry there is a natural pairing of adjacent spin-types to form three superstring fields or superfields. The following summarizes these correspondences (also refer to Figure 40):

- The first correspondence matches the merging of akasha and vayu to create vata dosha with the unification of the spin-2 graviton and spin-3/2 gravitino to form the gravity superfield. This is not only an exact numerical correspondence, but the qualitative characteristics of the gravity superfield also match the nature of the vata dosha. Qualities often attributed to vata dosha are movement and space. This is almost identical with nature of the gravity superfield responsible for space-time curvature and gravitational attraction of masses.

- The second correspondence matches the merging of agni and jala to create pitta dosha with the unification of the spin-1 force fields and spin-1/2 gauginos to form the gauge superfields. Gauge superfields are responsible for forces such as electromagnetism, the weak force and strong force. The weak force underlies the process of energy production in stars, radioactive decay and neutrino interactions. The strong force acts over short distances to hold together the structure of the atomic nucleus. Collectively, the nature of these forces corresponds to the nature of the pitta dosha which exhibits qualities of metabolism, heat, intensity and transformation.

- The third correspondence matches the merging of jala and prithivi to create kapha dosha with the unification of spin-1/2 matter fields and the spin-0 Higgs fields to create the matter superfields. Attributes associated with kapha dosha are cohesion, stability and structure. This closely corresponds to matter superfields responsible for the physical masses and matter of our universe.

In conclusion, we have described the sequential emergence from the unified field of the five fundamental spin-types and the three superfields. All elementary particles and forces can be analyzed and categorized from these basic entities. Because of the extremely small space-time distance scales of quantum physics, there is often little experimental evidence available to support the mathematical formulations of quantum field theories. For this reason, much of the research in this area has been advanced through speculations, ideas and mathematical notions. Nevertheless, even with the lack of complete experimental confirmation, we see a striking correspondence between the fundamental divisions of matter and energy as described by quantum physics and the pattern responsible for the genesis of the material universe recorded in the ancient Vedic Literature.

CHAPTER 11

CHEMISTRY: THE FIVE FUNDAMENTAL CLASSIFICATIONS MERGING TO FORM THE THREE PRIMARY PARTICLES OF AN ATOM

The language of chemistry, or the study of the structure and interconversions of matter, begins with the atom. Advancements in our understanding of atomic structure during the past century have culminated in the discovery of a vast array of substances called elements which make up all matter in our universe. Earlier in this book we discussed the periodic table of elements, the basic principles of atomic structure, and demonstrated that the basic structure of an atom exactly matched the numerical structure of the Constitution of the Universe. In this chapter, we build upon this knowledge to reveal yet another striking parallel between our modern understanding of chemistry and the wisdom of the ancient Vedic civilization.

The Atom (Samhita) is the Building Block of Elements

Figure 41 presents an overview of the correspondences to be discussed in this chapter between the fundamental building blocks of atomic structure and the pattern by which the five tanmantras or primal entities described in the Rig Veda begin to create the material universe. At the bottom of this figure we have the atom (from Greek "atomos" meaning indivisible) which is considered to be the smallest unit of an element that can exist and still retain its original chemical and physical properties. Further division of matter would destroy the atom and convert it to subatomic particles and energy. All atoms of a given element are identical, while compounds and

Figure 41. Genesis of the Atomic Structure
According to Chemistry: The Three Primary Particles
of an Atom Arising from Five Fundamental Classifications

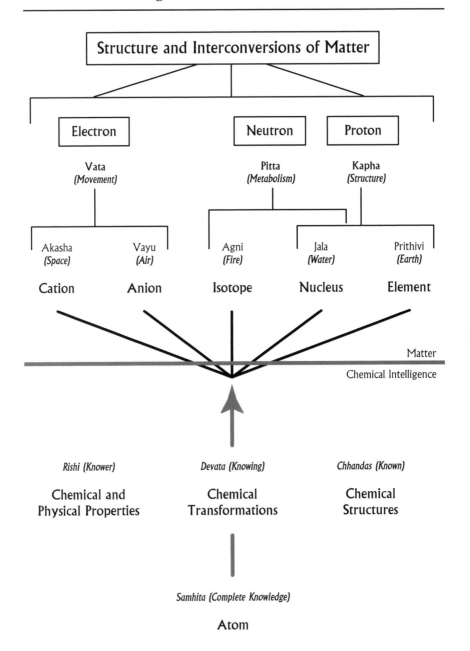

mixtures are made up of two or more types of atoms. Although all atoms are created from the same components, the number and arrangement of these components in an atom distinguishes one element from all the others. The elements of the periodic table of chemistry represent the building blocks that compose all matter in the universe. In relation to our understanding from Rig Veda, the atom and its structure correspond to the samhita value of intelligence (complete knowledge) because the language of chemistry, or the study of matter, really begins and is derived from the nature of atoms.

There are Three Classifications of Matter

A complete description and understanding of the nature of atoms can be obtained by characterizing matter in three ways. These three classifications are not separate, but rather each provides different information regarding the nature of the atoms constituting the elements of the periodic table of chemistry. Together these three categories correspond closely with the three aspects of samhita (complete knowledge): rishi (knower), devata (process of knowing) and chhandas (known). According to Rig Veda, these three aspects of knowledge or intelligence are simply intellectual conceptions or interpretations that provide a complete understanding of an entity.

The first classification of matter is the chemical and physical properties of an atom. Physical properties are the characteristics that can be identified without changing the chemical nature of the substance (e.g., boiling point, density, color, physical state and solubility). Chemical properties of a substance are observed when it undergoes a chemical change, i.e., transforming it into one or more other substances. An example of a chemical change is the ability of the element hydrogen and the element oxygen to react to form water. Together chemical and physical properties characterize or identify the nature of an element. These aspects of matter correspond in Rig Veda to the qualities of rishi (knower) which is associated with the inherent nature and abstract information of an entity (refer to Figure 41).

The second classification of matter, the chemical transformations of atoms, are changes in which one or more kinds of substances undergo a rearrangement of atoms to produce a new kind of matter or several new kinds of matter. Chemical changes include such diverse reactions as a seed sprouting and growing into a tree; a burning log which forms smoke and ashes; and the burning of chlorine gas and sodium metal to produce sodium chloride (edible table salt). Information regarding reaction mechanisms, chemical bonding and chemical kinetics is included in chemical changes. Since all these processes involve change and transformation, it makes sense to match the category of chemical transformations with the dynamic and active nature of devata, or the process of knowing (refer to Figure 41).

The last classification of matter, the chemical structure of atoms, portrays information on the shape and distribution of atomic components. For example, the molecular geometry of an atom or molecule includes knowledge of bond angles, orientation of other atoms around a central atom, and the shape of atomic orbitals. Other information contained in chemical structure includes the arrangement and distribution of atomic components within the orbitals and suborbitals. This information on the chemical structure of atoms precisely matches the qualities in Rig Veda of chhandas (known) which is associated with the material nature and characteristics of an entity (refer to Figure 41).

Collectively, these three general classifications of matter provide information regarding the nature of atoms and the substances making up our universe. The chemical properties, transformations and structure of a particular substance are, however, determined from the structural arrangement of the components within each atom. Historically, the early research for these smaller components (subatomic particles) from which atoms are built was largely pioneered by three physicists: J.J. Thomson, Ernest Rutherford and Robert A. Millikan. The decades of research performed by these three scientists culminated in the foundational concepts governing the modern view of atomic structure.

Atoms are Composed of Electrons, Neutrons and Protons

From the simplest point of view, an atom consists of three major kinds of particles: the electron, the neutron and the proton. The nucleus or center of the atom contains most of an atom's mass, yet occupies only a small portion of the atomic volume. Contained within the atomic nucleus are positively charged protons and neutrons that have no charge, yet have the same mass as protons. Revolving around the nucleus are electrons carrying a fixed negative charge, equal in magnitude to the positive charge of a proton, and having a mass about 2,000 times less than either the proton or the neutron. The radius of an atom is about 10^{-8} cm compared with the nuclear radius which is approximately 10^{-13} cm. One way to conceptualize the structure of an atom is to think about it in macroscopic terms. For example, if we were to imagine an atom to represent a huge ball of empty space nested in a football stadium, the nucleus would appear to be the size of an orange, and the electrons would seem like pin heads revolving around the nucleus at high speeds. Visualizing an atom like this enables us to realize that the volume of an atom is almost entirely empty space.

Each element contains a specific number of electrons, protons and neutrons, which make it distinct and unique. The number of protons in the atomic nucleus is called the atomic number and is used to identify the name of an element. Simply put, if the proton number changes, then the name of the element changes. For example, an atom with one proton refers to the element hydrogen, whereas an atom with eight protons refers to the element oxygen. In the periodic table used by chemists today the elements are arranged according to the atomic number in ascending order.

The number of neutrons in the atomic nucleus is called the neutron number. Added together, the number of protons and the number of neutrons equal what is called the mass number. From the mass number the overall mass of the atom is calculated (normally the mass of electrons are not included in this calculation because their contribution is insignificant). The number of neutrons in an atom may vary independently of the number of protons and electrons. However, since neutrons are electrically neutral (no charge) the net charge of an atom will be unaffected by the gain or loss of neutrons. Atoms containing an unequal number of neutrons and protons are called isotopes. Because the chemistry of an atom is largely due only to its electrons (refer to Chapter 3, Section I), isotopes usually exhibit the same chemical properties as compared to atoms which have the same number of protons and neutrons. Since many isotopes are radioactive, the nuclei of these atoms are unstable and over time will undergo a process known as radioactive decay. Radioactive decay is a nuclear reaction in which the atomic nucleus spontaneously disintegrates, giving off subatomic particles and energy, until it breaks down or decays into atoms that are more stable. Of the approximately 2,000 known nuclides (a nuclide is a common term applied to each unique atom), only 279 are stable with respect to radioactive decay. All elements on the periodic table that have eighty-four or more protons are subject to radioactive decay. Also, every known element has at least one isotope which is radioactive. In nature, elements are usually found as a mixture of isotopes, and many have been isolated and/or synthesized for research and technological purposes.

Elements are usually electrically neutral because the number of electrons is equal to the number of protons. Although the number of protons in an atom generally remains fixed, outside perturbations in the environment can cause one or more electrons to be pulled away from an atom, or added to it. When an atom gains or loses electrons, it ends up being negatively or positively charged, respectively. Gain or loss of electrons, however, does not change the name of the element. Electrically charged atoms are called ions and are divided into two groups: negatively charged ions which have gained one or more electrons, called anions; and, positively charged ions which have lost one or more electrons, called cations. Although ions make up less than 1 percent of the mass of living matter, they play a particularly crucial role in areas such as biological activity. The number of electrons in an atom can change because elements tend (with the exception of noble gases) to "want" to gain or lose electrons to achieve the maximum number of electrons in their outermost atomic orbital. As a general rule, atoms that have completely filled outer energy levels are the most stable and unreactive. A good example of the formation of ions is the reaction between neutral chlorine and sodium which creates common table salt. In this reaction one electron is stripped from the sodium and is added to the chlorine. This creates a sodium cation with a net positive charge of 1 and a chlorine anion with a net negative charge of -1. Because anions and cations have opposite charges, they attract each other. In the reaction just described this is indeed what happens, the sodium cation and chlorine anion unite to form solid sodium chloride (table salt). Sodium loses an electron and chlorine gains an electron because this one loss/gain process allows each atom to achieve the maximum number of electrons in its outermost orbital.

The Five Tanmantras Correspond to Five Aspects of Atomic Structure

At this stage, we have provided sufficient background information to draw a complete correspondence between the basic principles of atomic structure and the pattern in which the five tanmantras or primal entities described in the Rig Veda begin to create the material universe. According to the Vedic literature, these five tanmantras or fundamental categories of matter and energy arise from a universal field of consciousness or intelligence. Similarly, there are five basic classifications or variations arising from the structure of an atom. Figure 41 presents these five basic variations of the atomic structure and their corresponding tanmantras. The following is a description of each correspondence:

- The first two variations of the atomic structure are called the cation and the anion. Cations are created by decreasing the number of electrons, and anions by increasing the number of electrons in their outermost atomic orbitals. Because changes in the number of electrons govern the formation of both kinds of ions, it is appropriate to link these two variations of the intra-atomic arrangement with the first two tanmantras, akasha (space) and vayu (air). Electrons, as described earlier, have a relatively small mass, to the extent that they are often described as possessing a dual nature, that is, they have both a wave-like and a particulate nature. As a result, it is experimentally impossible to know simultaneously both the momentum and position of an electron at any given time. For these reasons electrons act as though they are spread out over large spaces forming "diffuse clouds" which tell us the probable places where we might locate them. Another important feature of the atomic structure is that the diameter of the nucleus is about five orders of magnitude smaller than the diameter of the entire atom. Thus, most of the atomic volume is nothing more than empty space with minute electrons whirling around a dense nucleus at lightning speed. These properties of an electron correspond to characteristics embodied by the tanmantras akasha (translated as space) and vayu (translated as air). For these reasons cations and anions, which are defined as changes in the number of electrons, are associated with these two tanmantras.

- The third variation of atomic structure, called an isotope, involves changes in the number of neutrons in the nucleus. The name isotope (from Greek "iso" meaning equal, and "topos" meaning place or position) was originally proposed to denote a difference of masses in the same element caused by changes in neutron number. As already discussed, a large number of isotopes exhibit radioactive decay which may occur in various ways. Among these the more important are called alpha, beta and gamma emission. Alpha radiation involves nuclear decomposition that produces α particles (identical with the helium nucleus, that is, two protons and two neutrons). Beta radiation involves the production of ß particles or the flow of electrons. Gamma radiation, emission of electromagnetic radiation, is basically different from alpha and beta radiation which are particulate. Because of the radioactive nature (hence instability) of many isotopes, it was clear that isotopes correspond best with the agni tanmantra (translated as fire). It is also interesting to note that in scientific research radioactive substances are labeled as "hot" and are stored in "hot rooms," which matches the fire nature inherent in the agni tanmantra. An example of a radioactive isotope is Uranium-238 (containing ninety-two protons and 146 neutrons) that has been used in research on nuclear fuel.

• The fourth variation of atomic structure, the nucleus of an atom, can be viewed as a collection of protons and neutrons. As described earlier the atomic nucleus exhibits several impressive characteristics, most notably, its very large density and its extremely small size. To help conceptualize just how dense the nucleus is, it may be helpful to compare it with a macroscopic object. For instance, if we had a sphere the size of a ping-pong ball composed entirely of nuclear material, it would have a mass of 2.5 billion tons. These characteristics suggest that the nucleus may correspond to the jala tanmantra (translated as water). The small spherical and dense nature of the nucleus is suggestive of the forces of cohesion which act to form a "water-drop." It may be of interest to note that a similar interpretation of the jala (water) tanmantra is described in Chapter 16 for one of the Platonic Solids (the icosahedron). The strongest evidence, however, for associating the nucleus with the jala tanmantra will become evident shortly when we discuss how the five tanmantras combine to form the three doshas.

• The fifth variation of atomic structure is the electrically neutral element. In a neutral element the number of protons, neutrons and electrons are all equal to one another. The name of an element and its general chemical/physical properties are determined by the number of protons in its atomic nucleus. If the number of protons in an element changes, then the element, and therefore the name, also changes. There are ninety-two elements occurring naturally on Earth, and at least twenty-four are essential to living matter. For these reasons the neutral element is associated with the prithivi tanmantra (translated as earth).

The Electron, Neutron and Proton Match the Creation of the Three Doshas

According to the Rig Veda, after the five tanmantras or primal entities have emerged from the universal field of consciousness, they merge together in a well-defined pattern to create three new entities called doshas. The way they merge is as follows: akasha (space) and vayu (air) tanmantras merge to form vata dosha; agni (fire) and jala (water) tanmantras merge to create pitta dosha; and jala (water) and prithivi (earth) tanmantras merge to produce kapha dosha. Likewise, the five fundamental variations of the atomic structure are grouped according to the three primary building blocks of an atom: the electron, the neutron and the proton. The following summarizes these correspondences:

- The first correspondence matches the merging of akasha and vayu to create vata dosha with the grouping of cations and anions according to changes in the number of electrons. Not only is this an exact numerical correspondence, but also the qualitative characteristics of an electron match very well with the nature of the vata dosha. Qualities often attributed to vata dosha are movement, lightness, smallness and unpredictable like the changing wind. This is almost identical to an electron which is very minute, has low mass and travels at lightning speed through the virtually empty space of an atom. In addition, because the electron exhibits both wave-like and particulate nature it is impossible to predict both its location and momentum simultaneously.

- The second correspondence matches the merging of agni and jala to create pitta dosha with the grouping of isotope and nucleus according to changes in the number of neutrons. Both isotopes and the composition of the atomic nucleus are determined by the number of neutrons in an atom. A change in neutron number alters the composition of an atom's nucleus and thereby creates an isotope. As described earlier, the qualities associated with radioactive isotopes correspond to the attributes of agni tanmantra (translated as fire). In addition, some of the qualities of pitta dosha (fiery, hot and scorching) also match radioactive isotopes which are created as a result of changes in the number of neutrons.

- The third correspondence matches the merging of jala and prithivi to create kapha dosha with the grouping of nucleus and element according to changes in the number of protons. Both the identity of an element and the composition of the atomic nucleus are governed by the number of protons in an atom. However, the composition of the atomic nucleus includes protons as well as neutrons. This is why the placement of the nucleus in Figure 41 is so critical. As alluded to earlier, the nucleus corresponds to the jala tanmantra (water) because it is composed of both neutrons and protons, which is necessary if these components of atomic structure are to match exactly the pairing sequence of the tanmantras. In addition to these exact numerical correspondences, the properties of kapha dosha are similar to those exhibited by the proton. Qualities associated with kapha dosha are structure, cohesion and stability. Correspondingly, the proton is responsible for an element's identity and, in part, for the formation of the dense, solid and stationary atomic nucleus.

In conclusion, we have presented a definitive correspondence between the fundamental building blocks of atomic structure as described by modern chemistry and the ancient Vedic understanding of how the five tanmantras begin to create the material universe. From these basic variations in atomic structure all elements are created. These elements in turn provide the basic structures for all matter and its interconversions in our universe.

CHAPTER 12:

BIOLOGY: THE FIVE NUCLEOTIDE BASES MERGING TO FORM THE THREE BASE-PAIRS

Throughout the range of biological life, DNA is the fundamental basis and underlying source of all physiological structure and activity. Although DNA itself is not biological information, it is the medium through which biological information is stored, transmitted and used to direct all aspects of cellular functioning. Every portion of the DNA molecule contains information to create specific proteins which at increasingly higher levels of physiological organization interact with one another to create a holistic organism. Earlier in this book we discussed the basic principles of DNA and demonstrated that the structure of the genetic code exactly matched the numerical structure of the Constitution of the Universe. In this chapter we build upon this knowledge and reveal yet another striking correspondence between our modern understanding of DNA and ancient Vedic wisdom. This new correspondence demonstrates that the mechanics of DNA expression precisely matches the numerical pattern responsible for the genesis of the material universe as understood by the Vedic tradition.

The Totality of Genetic Information is Represented in the Genome (Samhita)

Figure 42 presents an overview and a summary of the parallels between the expression of genetic information and the structural mechanics responsible for the genesis of the material universe to be discussed in this chapter. At the bottom of this figure we have what is called the "genome," which represents all the DNA residing in a cell. The genome is the totality of genetic information or the complete set of blueprints specifying every function and structure of a cell. It is responsible for telling the body how

Figure 42. Genesis of DNA According to Biology:
The Three Base-Pairs Arising from the Five Nucleotide Bases

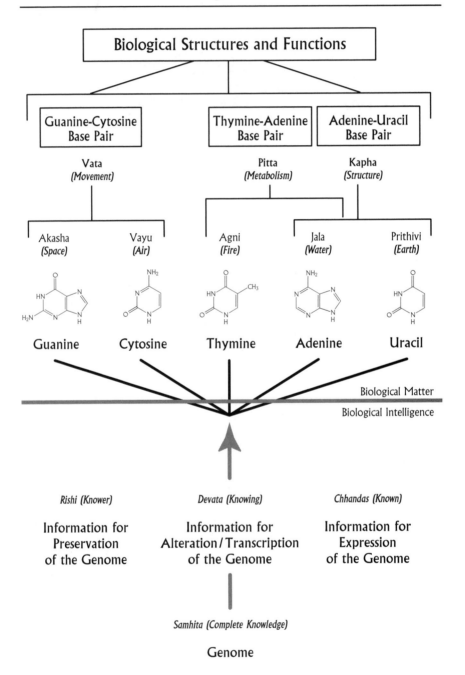

to grow, maintain and reproduce itself. In relation to the Vedic understanding of the genesis of the material universe, the genome corresponds to the samhita value of knowledge which represents the summation of rishi, devata and chhandas.

The totality of genetic information can be grouped into three general categories that correspond with the flavors of rishi (knower), devata (process of knowing) and chhandas (known). The first division of genetic information is the information responsible for the preservation of the genome. Biochemical mechanisms mediating this type of information include DNA replication and repair. This division corresponds with the nature of rishi which is associated with knowledge itself or abstract information. The second division of genetic information is for the alteration and transcription of the genome. Chemical mechanisms involved in this type of information include DNA recombination, rearrangement and mutations. Since all these processes involve some change or transformation, it is logical to match this type of information with the dynamic and active nature of devata. The last division of genetic information is the information responsible for the expression of the genome. Physiological mechanisms involved with this type of information include DNA transcription and translation which create proteins needed in all structural and functional aspects of the cell. This matches the material nature of chhandas (Wallace, Fagan and Pasco, 1988).

The Five Nitrogenous Bases of DNA Correspond to the Five Tanmantras

To understand how genetic information expresses itself into biological life, we need to review the structural composition of DNA. DNA is made up of a large number of subunits, called nucleotides, which combine together linearly to form two very long threadlike chains coiled around each other to give what looks like a double-helix structure. Each subunit in DNA is composed of three parts: a nitrogen-containing base, a sugar and a phosphate group. The nitrogenous bases of DNA carry the genetic information, whereas the other two components serve a structural role.

Nitrogen-containing bases are the individual letters or building blocks of the language of life enabling genetic information to be stored, transmitted and used. There are exactly five nitrogenous bases involved in the expression of DNA: guanine, cytosine, thymine, adenine and uracil. Adenine and guanine are from the purine class of molecules, and thymine, cytosine and uracil are from pyrimidine class of molecules. Four of these nitrogenous bases occur in the DNA molecule itself (adenine, thymine, cytosine and guanine). The fifth nitrogenous base is found in the RNA molecules that are involved in the transcription and translation of genetic information. Each RNA molecule contains adenine, cytosine, guanine and uracil. Thus, the difference between DNA and RNA in base composition is that in RNA thymine is replaced by uracil.

As described earlier in the book, a precise sequence of the bases in DNA is used to specify how to create a protein. Each protein is made up of a series of amino acids linked together. The sequence or arrangement of these amino acids is determined by a corresponding sequence of bases in DNA. There are, however, twenty naturally occurring amino acids and only four nitrogenous bases in the structure of DNA. Thus, to have enough bases to code for all amino acids, the bases combine together into sets of three. A set of three consecutive bases, called a codon, specifies only one amino acid. The knowledge of which amino acid is specified by which codon is called the genetic code. By understanding the sequence of codons in DNA, we can determine the matching sequence of amino acids that constitute a protein; proteins in turn provide the building blocks of the body.

At this stage we can draw our first major correspondence between the expression of genetic information and the structure responsible for genesis of the material universe. According to the Vedic tradition, consciousness or the unified field of intelligence lies at the basis of creation. From this field the five fundamental categories of matter and energy called the tanmantras emerge. Similarly, exactly five nucleotide bases that are responsible for the preservation and expression of DNA arise from the totality or storehouse of genetic information in the genome. These five nitrogenous bases represent the material expression of biological information contained in the genome. In Figure 42 we have presented the five bases, their structural formulas and their corresponding tanmantras. The reason the five bases were placed in this particular order will become evident shortly.

The Three Types of Base-Pairing Match the Creation of the Three Doshas

To understand fully how the genetic information contained in the sequence of DNA bases is converted into proteins, we need to introduce another concept called "complementary base-pairing," term used to describe the chemical interactions between nitrogenous bases that are responsible for a number of important events. There are well-defined ways in which the five nitrogenous bases interact with one another. Guanine will interact or pair only with cytosine; thymine always interacts or pairs with adenine (in DNA); and, uracil will only interact or pair with adenine (in RNA).

In DNA adenine-thymine and guanine-cytosine are the two types of base-pairs responsible for holding the two strands of the DNA double-helix together. They act as rungs connecting together the two long chains. Thus, all the bases in one strand of the DNA molecule are complementary to the bases on the other strand. By knowing the order or sequence of bases on one strand, we can determine the order on the other strand using complementary base-pairing rules (i.e., guanine pairs always with cytosine, and thymine pairs always with adenine). This structural facet of the DNA molecule immediately implies a mechanism for DNA replication and conservation of genetic information from one cell generation to the next. When a cell divides, the DNA molecule "unwinds" and the two strands separate into the two daughter cells. In each daughter cell the separate strands of DNA act as a pair of templates, each parent strand remaining intact as a new companion strand is assembled on each one by complementary base-pairing. As a result, both daughter cells have the same DNA duplicated exactly. Through this process genetic information can be passed from one cell generation to another and remain unaltered.

Using this same process of complementary base-pairing, genetic information is converted from a DNA template into a protein. For this to happen a sequence of biochemical processes occur which are divided into two general categories; transcription and translation. Transcription involves the transfer of genetic information from a DNA template to an mRNA molecule. Translation involves the conversion of genetic information in mRNA into the sequence of amino acids in a protein. Both these processes are possible because of complementary base-pairing. Refer to Chapter 4, Section I for more details. Once a protein is synthesized, it can be used as a building block according to the needs of the organism.

Based on this general overview of DNA expression, we can draw our second and final correspondence with the primal patterns responsible for the genesis of the material universe. According to Vedic understanding (refer to Chapter 9), after the five tanmantras have emerged from the unified field of intelligence, they merge together in a well-defined pattern to create three new entities called doshas. The way they merge is as follows: akasha and vayu tanmantras merge to form vata dosha; agni and jala tanmantras merge to create pitta dosha; and, jala and prithivi tanmantras merge to produce kapha dosha. Likewise, the five nitrogenous bases in DNA combine together according to well-defined rules of complementary base-pairing which are essential for DNA replication, transcription and translation. Further, the bases combine together according to base-pairing rules in exactly the same order as the tanmantras merging to form the three doshas. The three base-pairs are as follows: guanine-cytosine, thymine-adenine, and adenine-uracil. A summary of these striking parallels is shown in Figure 42.

In conclusion, we have presented a very clear correspondence between the expression of genetic information in DNA and the genesis of the material universe as understood by the ancient Vedic civilizations. Through the expression of DNA all biological structures and functions are specified which are responsible for the growth, reproduction, metabolism and development of all organisms.

CHAPTER *13*

PHYSIOLOGY: THE FIVE SENSES MERGING TO FORM THE THREE TYPES OF RECEPTION

Sensory modalities are the primary link between our nervous system and the events occurring within and around us. Typically our sensory modalities are divided into five basic groups – hearing, touch, sight, taste and smell – each one providing a unique type of information from environmental stimuli. An environmental stimulus results from any change occurring in the surroundings – if there were no changes in our universe, then the world would appear perpetually devoid of details. When an environmental stimulus is "sensed" or "perceived," the information is transformed into electrochemical messages that can be processed by the nervous system. Through such mechanisms our five senses allow us to adapt, survive, cope and interact with the world around us. They provide the building blocks that allow us to construct our conceptual picture of the universe. In this chapter we will discuss the dynamics of the five senses and demonstrate that they match the numerical structure responsible for the genesis of the material universe as understood in ancient Vedic tradition.

Common Sense (Samhita) and the Three Aspects of Sensory Modalities

In Figure 43 we have presented an overview and summary of the parallels between the dynamics of the five senses and the structural mechanics responsible for the genesis of the material universe which will be discussed in this chapter. At the bottom of this figure we have what is known as "common sense," the net result of sensory perception. Common sense is our innate and inborn ability to make good judgments and decisions. It allows us to interact and live in the world by making choices that

Figure 43. Genesis of Sensory Modalities According
to Physiology: The Three Types of Reception
Arising from the Five Senses

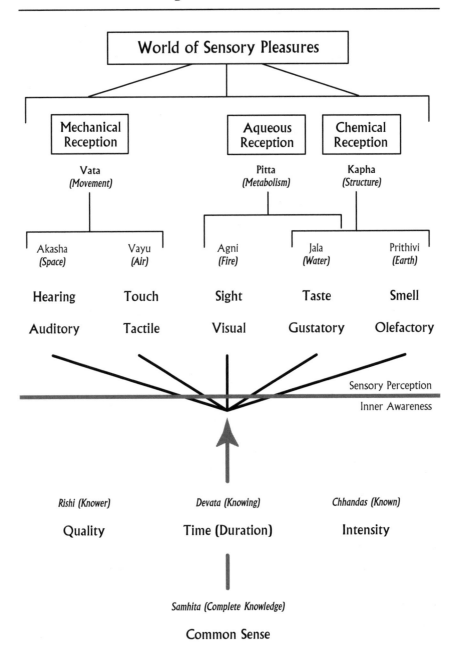

are reasonable, logical and progressive. In relation to the Vedic under-standing of the genesis of the material universe, common sense corre-sponds to the samhita value of knowledge which represents the togetherness of rishi, devata and chhandas.

However, for common sense to function properly, the five individual senses need to tell common sense what to do. Each of the five senses can be thought of as five primary branches of common sense. Together these five senses contribute information which guides the holistic and balanced func-tioning of common sense. All five senses can be described in three distinct ways – intensity, quality and time (Marks, 1978). These three aspects of sensory modalities correspond to the flavor of rishi (knower), devata (process of knowing) and chhandas (known). The first characteristic is the "quality" of the sense. For example, when we touch something it could be soft or hard; when we taste something it is pungent, astringent, sour, sweet, salty or bitter; when we see something it is a particular color such as blue or red. This aspect of a sense corresponds with rishi which is equated with the inherent or abstract nature of an entity. The second characteristic, the "time duration" of a sense, refers to the length of time that a sense is exposed to an environmental stimulus. This corresponds to the nature of devata which is associated with movement, duration and dynamism. The third characteristic, the "intensity" of a sense, describes whether a stimulus is barely detectable or is overwhelming and unpleasant. For example, a sound may be a whisper which caresses the ear or the blaring of an ambu-lance; a light may be a flicker in the corner of our eye or the dazzling rays from the sun. This is equated with the chhandas interpretation which expresses the physical or material nature of a sense.

The Five Senses Correspond to the Five Tanmantras

General characteristics of each of the five senses will now be described. Following each description we will match that particular sense with one of the tanmantras. We were inspired to draw this correspondence between the five senses and the five tanmantras by the classical texts of Ayurveda (the ancient medical system of Vedic India). In these texts (Caraka, Sutrasthana, VIII.14 and IX.8-10) the five tanmantras are discussed in relation to healing and are also clearly matched to the five senses. These exact correspondences between the tanmantras and the five senses are used in this chapter.

- The sense of hearing is responsible for the detection of sound. Sound waves consist of mechanical vibrations which periodically displace atoms and molecules until they reach the receptors in our ears. This explains why sound cannot be transmitted in a vacuum because there are no atoms or molecules to be displaced. The organ in our body responsible for hearing is the ear which acts to funnel and collect sound waves (this is especially true of the ears of bats, donkeys and hares). These concentrated sound waves then become translated into electrochemical signals that are sent to the nervous system and processed. Since sound waves involve the mechanical vibration of molecules and have spatial characteristics, Ayurveda associates hearing with the Akasha or Space tanmantra.

- The sense of touch is considered a general sense that has no specialized organ. It is a "cutaneous sense," meaning that the sense of touch is scattered throughout all parts of our skin. This sense is one of the most complicated senses because a variety of external stimuli influence it (e.g., pressure, cold, heat and pain). To detect each of these stimuli, they must physically move through air to touch us. For this reason Ayurveda associates the sense of touch with the Vayu or Air tanmantra. (Note: we can still touch things when submerged in water, but we are much less sensitive. For example, it is very easy to receive a cut in water and not know until we come out.)

- The sense of sight is responsible for vision and seeing objects. Our eye is able to see things because of photoreceptors that are able to detect and amplify light to detect things clearly in the environment. These receptors are similar to the lens a child might use to concentrate sunlight and set a piece of paper on fire. According to Ayurveda, the sense of sight corresponds to Agni or Fire tanmantra. One of the qualities of fire is to illuminate objects so that we can see them.

- The sense of taste is responsible for savoring the food that we eat. Taste buds on our tongue evoke the sensation of taste, such as, sour, salty, bitter or sweet. However, to taste something, the food or ingested substances must first be dissolved in the fluid that bathes the tongue and interacts with the taste buds. A dry tongue will not taste anything. The saliva in our mouth is water which allows food to be tasted. For this reason, the Ayurvedic texts associated the sense of taste with Jala or Water tanmantra.

- The sense of smell through our nose is responsible for smelling things. We smell volatile chemicals that dissolve in the mucous coating of the nasal cavity. Because this sense operates through the interaction with physical chemicals, Ayurveda associates this sense with the Prithivi or Earth tanmantra.

The Creation of the Three Doshas Matches the Three Types of Reception

We have now made a clear correspondence between the five tanmantras and the five senses. Although the classical texts of Ayurveda match the five tanmantras to the five senses, this is as far as the correspondence goes. There are, however, well-defined characteristics shared by the senses which would allow them to merge into three general categories. This allows us to draw our complete correspondence between the dynamics of the five senses and the structure responsible for the genesis of the material universe.

The clearest example of senses exhibiting similar characteristics is the sense of smell and the sense of taste. Often these two senses have been called the "chemical senses" or "contact senses" (Starr and Taggart, 1984). Both taste and smell require physical molecules and contact to evoke a sensation. Also, because the nasal cavity is open to the throat, vapors from food in the mouth easily move into the nose. Hence, much of what we consider to be taste is actually the sensation of smell. This is particularly evident when we have a cold and all we experience when we eat is the unpleasant sensation of pure taste alone. This close relationship between smell and taste prompted us to merge these two senses into one division called "chemical reception." By chemical reception we mean that both the sense of smell and the sense of taste receive environmental stimuli in the form of physical or chemical molecules. This matches the same merging of Jala and Prithivi tanmantras to create kapha dosha as described in the ancient Vedic Literature (refer to Figure 43).

The second general category of senses having similar characteristics involves the sense of taste and the sense of sight. Both these senses require an aqueous or watery medium to detect environmental stimuli and to function properly. As mentioned earlier, the sense of taste requires the presence of saliva to dissolve the food. A completely dry tongue will not evoke the taste sensation when food is eaten. To conceptualize this, think of eating dry, unsalted bread or crackers – they do not have much taste. For the sense of sight we need to comprehend how our eyes are made and how they function. Our eyes consist of a huge number of photoreceptors whose purpose is to focus or collect light (photons) so that we can "see." Conceptually, the simplest type of photoreceptor would only detect light. A more sensitive device which could amplify light from the environment would function as follows. Light from the air would pass into a curved surface of water which would tend to bend the rays of light towards the center, regardless of where they would strike. The net result would be to collect or focus incoming light to allow for better vision. This more sensitive photoreceptor is exactly how our eyes are designed, but instead of using water itself, our eyes are basically transparent objects that are largely filled with a watery-fluid (Asimov, 1963). This fluid is under substantial internal pressure (about 177 mm Hg higher than external air pressure) giving the eye the shape of a rigid sphere. Also, the eyes produce tears which are used to protect against foreign matter (e.g., smoke and chemicals in the air).

In addition to these two sensations sharing similar characteristics, the sense of sight can evoke the sense of taste. For instance, the sight of a food, such as a lemon, will excite the sensation of taste which will cause us to salivate. Because of these relationships between sight and taste, and the fact that both senses fundamentally require the medium of water to function properly, we propose that they comprise a second division called "aqueous reception." By aqueous reception we mean that both the senses of sight and taste receive environmental stimuli (photons or chemical molecules) through some watery medium. This matches the same merging sequence as Agni and Jala tanmantras which form the pitta dosha as described in the genesis of the material universe.

The last general category of senses having similar characteristics involves the sense of hearing and the sense of touch. Together these two senses are often called the "mechanical" senses because they involve the mechanical vibration of molecules (Starr and Taggart, 1984). For the sense of hearing, sound waves produce periodic pressures on anything they contact. Although sound can travel in any medium, air is one of the most common. For the sense of touch, we feel this sensation when some object or air molecules physically touch us. This happens mostly through the medium of air. As mentioned earlier, this can also take place in water, but our sensitivity is dramatically decreased. Along these lines, research has shown that when subjects are given alternating auditory and cutaneous stimuli, they often report that a weak auditory click resembles a cutaneous tap and a pulse on the fingertip resembles a sound (Gescheider and Niblette, 1967). It has also been shown that the intensity of a touch stimulus may be significantly reduced by simultaneous auditory stimulation. The reverse is also true. These relationships between the sense of touch and the sense of hearing prompted us to establish a third division called "mechanical reception." By mechanical reception we mean that both the senses of hearing and touch receive environmental stimuli through the mechanical distortion of sound waves and tactile stimulation. Mechanical reception corresponds to the merging of Akasha and Vayu tanmantras which produces the vata dosha as described in the mechanics responsible for the genesis of the material universe.

Although there are many other possible similarities between the senses, we feel that these three are the most important. Perhaps more research in this area will continue to give support to the value of these three divisions which are already known and generally accepted.

In conclusion, we have presented an interesting correspondence between the dynamics of the five senses and the pattern responsible for the genesis of the material universe as understood by the ancient Vedic civilization. Through the constant interaction of the five senses we develop the proper common sense which enables us to enjoy optimally the world of sensory pleasures.

CHAPTER 14

MUSIC: THE FIFTH MERGING TO FORM THE MAJOR TRIAD

Since time immemorial music has existed in every major culture and has been upheld for its universal attraction, beauty and power. Although musicians have intuitively understood how to construct and play music to create pleasure and repose, they have generally avoided the objective understanding of the nature of musical sound. Consequently, most of our current knowledge of music theory is attributed to the work of scientists. Earlier in this book we discussed some of the basic principles of music theory in relation to the Constitution of the Universe. In this Chapter we build on this information and reveal yet another correspondence between music theory and ancient Vedic wisdom. This new correspondence demonstrates the way fundamental musical intervals and chords are constructed matches the numerical pattern responsible for the creation of the material universe as understood by the Vedic tradition.

A Musical Tone (Samhita) is Characterized by Quality, Pitch, and Force

Figure 44 presents an overview and a summary of the parallels between the basic principles of music theory and the structural mechanics responsible for the genesis of the material universe. At the bottom of this figure we have placed the most fundamental aspect of music, a musical tone. Musical tones are distinguished from other sounds because they are composed of simple, regular and uniform sound waves which exert a rapid periodic function (Helmholtz, 1954). Contained within a musical tone are all tonal relationships and musical intervals in the structure of the harmonic overtone series. In relation to the Vedic understanding of the genesis of the material universe, a musical tone corresponds to the

Figure 44. Genesis of Musical Tones and Intervals: The Three Notes of the Major Triad Arising from the Musical Intervals of the Just Fifth

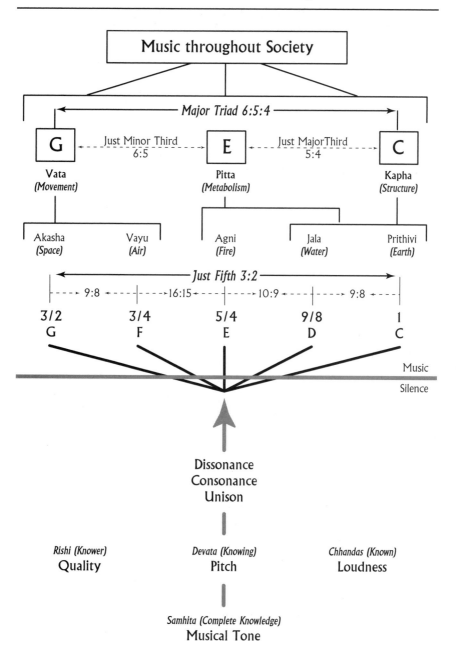

samhita value of knowledge and represents the togetherness of rishi, devata and chhandas.

Each musical tone is distinguished from other tones in three ways that correspond to the flavors of rishi (knower), devata (process of knowing) and chhandas (known). The first attribute describing a musical tone, called "quality," is the distinctive feature we hear between instruments that play sounds of the same volume and pitch. This aspect of a musical tone corresponds to the nature of rishi which is equated with the inherent or abstract nature of a tone. The second way to describe a musical tone, called "pitch," is simply the frequency or cycles/second of that sound. Frequency implies movement, dynamism and action which are exactly the qualities embodied by the devata interpretation. The final distinguishing quality of a musical tone is called "force," a measure of how loud a particular sound is. This quality is equated with the chhandas interpretation which expresses the physical or material nature of the tone. It is the force behind a tone that produces the actual physical sensation of a musical tone.

Description of the Basic Musical Intervals

Although a tone is considered to be a building block of music, by itself it does not create music. Pure isolated notes, like pure colors alone, create little if any interest or enjoyment for the listener. Music begins when one note is combined with another note. When two tones are combined, they are sounded together. Such a relationship between two tones is called a musical interval. Since a musical tone can be expressed as a numerical frequency, an interval is generally represented as a mathematical ratio that expresses the distance in pitch between the two tones.

Musical intervals are classified into two groups: consonance and dissonance. A consonant interval is the combination of two tones which when played together are generally agreed to give a feeling of completeness, satisfaction and pleasure. The opposite of consonance is dissonance. A dissonant interval is the combination of two tones which when played together give a feeling of harshness, clashing, incompleteness and restlessness. Typically a dissonant interval requires another interval to follow it to render a satisfactory effect (Backus, 1969; Helmholtz, 1954).

It is universally acknowledged that two musical tones sound well together when their frequencies can be expressed as the ratio of small whole numbers. The smaller the numbers in the interval ratio, the more consonant the interval. The further we proceed from small whole number intervals, the more dissonant the sounds become. This phenomenon was demonstrated by Pythagoras 2500 years ago using a monochord of one taut string divided into two adjustable, simultaneously vibrating segments. The question arises, however, why consonance is associated with the ratios of small numbers. Although many attempts have been made, this question has not been fully answered. The answer given by the ancient Greeks and ancient Chinese was that small numbers exhibited power and were the source of all perfection (Jeans, 1968).

The simplest conceivable whole numbered ratio is 1:1 which represents two musical tones of the same pitch (frequency) being played simultaneously. This musical interval is called a unison and represents two identical musical tones being played simultaneously. For example, when two singers sing at the same pitch, they are said to be in unison.

The next simplest whole numbered musical interval is 1:2 and is called the octave. An octave is the interval between the first note and the last note of a scale and is considered to be the most rudimentary and simplest. According to Helmholtz, the octave is called an "absolute consonance" because the prime tone of one of the combined tones is also some partial tone of the second note (Helmholtz, 1954). The mathematical significance of the perfect consonance of the octave is that the frequency of the second tone is double that of the first tone. Any two notes that are an octave apart are different, but because they exhibit a similar sound quality, they are given the same letter name. In fact, the resemblance of two tones an octave apart is so strong that, to the normal listener, they produce the effect of one sound. This special feature of the 1:2 relationship has often been termed the "basic miracle of music."

The next group of musical intervals include the fifth and fourth which are termed "perfect consonances" (Helmholtz, 1954). A fifth is the musical interval between the first and fifth tone in most scales (written mathematically as 2:3). One example of a fifth is when the tones C and G are sounded together on the Just or Natural Scale. It is possible, however, to start with any other pitch and construct the fifth interval. This interval is considered fundamental in music theory because of its use in the construction of scales and chords. For example, both the normal eight-note scale and the pentatonic scale are constructed using fifths. In the pentatonic scale only four sequential steps in the intervals of the fifth are required to generate a series of five notes. Continuation of these steps will generate the eight-note scale. Another important point regarding the fifth interval is the importance attributed to it by the ancient Greeks. Pythagoras considered the musical fifth to be the most beautiful and special interval. Not only was the fifth fundamental in ancient Greece, but also in Gregorian chant, Bach, Beethoven, Elvis Presley and others (Holtzman, 1994). The fourth is the musical interval between the first and fourth tone in most scales (written mathematically as 3:4). This interval is a less perfect consonance than the fifth. Its importance arises because it is the inverse of the fifth. This means that, when the fourth interval is combined with the fifth, it will make up one octave (i.e., 3/2 x 4/3 = 4 /2 or 2/1).

The last group of musical intervals we will describe are called the major third (4:5) and the minor third (5:6). These intervals are classified as being "medial" and "imperfect" consonances, respectively (Helmholtz, 1954). Both the major third and minor third are used in conjunction with the fifth in the formation of triads and other chords. An example of the minor third is when the notes E and G are sounded together on the Just Scale; an example of the major third is when the notes C and E are sounded together on the Just Scale.

From Unison to Consonance to Dissonance

Musical intervals that are more complex than the minor third become decreasingly consonant and pleasant to the listener. This is because in musical intervals which are large and complex the sound waves of the tones do not interact smoothly giving rise to unpleasant sounds. As a result, the interference between such sound waves creates rapid fluctuations in volume called beats. The outcome is called dissonance.

There are several ways to explain why small, whole numbered intervals are the most consonant. For example, in Figure 45 we have displayed the sine waves of seven different musical intervals. Both the x and y axis are drawn to the same scale on all graphs. Visually these sine wave curves allow us to compare the various musical intervals as they proceed from unison to consonance to dissonance:

- The first graph represents unison or two identical musical tones being played simultaneously. This musical relationship is drawn as a simple, smooth and continuous sine wave.

Figure 45. A Visual Representation of the Musical Intervals: From Unison to Consonance to Dissonance

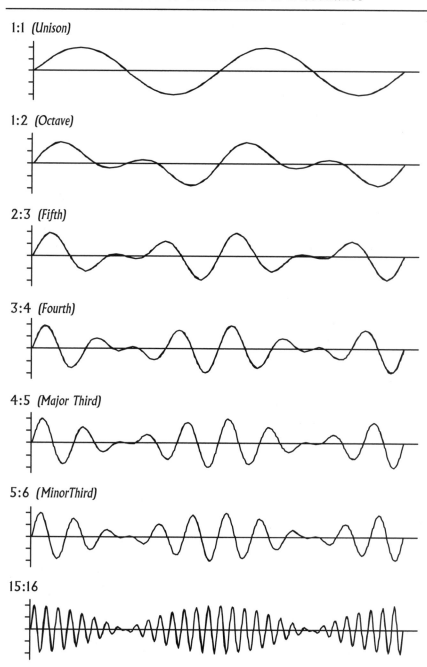

- The second graph depicts the octave interval. In this interval for every one vibration made by one tone, the other tone makes two vibrations. Figure 45 displays the resultant sine wave curve when two tones, which are an octave apart, are played together.

- The remaining graphs in Figure 45 represent other examples of musical intervals as they become more complex. Examining the range of sine wave patterns for each musical interval, we can derive several conclusions. First, as we progress from small to larger whole numbers, the sine wave patterns become increasingly complex and diversified. This increasing complexity is heard by the listener as increasing dissonance. Second, it is evident that as the musical interval becomes a ratio of larger numbers, the interaction of the two tones becomes "rough" and gives rise to rapid fluctuations in volume which are heard as beats. This is clearly depicted in the last graph in Figure 45 which displays the musical interval 15:16 where the sine wave curve looks like a pulse or a beat. Waves of this nature signal the presence of dissonance.

Another way to understand why small, whole-numbered intervals are consonant is to examine how many harmonics or partials are common to the two tones. It has been found that musical tones that sound good together have a lot of overtones in common. For example the interaction between the two tones of an octave can be represented as follows:

Fundamental	Overtones →						
Tone A: 100	200	300	**400**	500	**600**	700	**800**
Tone B: 200	**400**	**600**	**800**	1,000	1,200	1,400	1,600

Numbers in bold indicate those frequencies common to both tones. As readily observed, half of the partials of Tone A are also found in Tone B. Tones like these which share large numbers of overtones exhibit a high degree of consonance.

The next example represents the musical interval 8:9 which is considered to be a dissonant sound:

Fundamental	Overtones →						
Tone A: 800	1,600	2,400	3,200	4,000	4,800	5,600	6,400 7,200
Tone B: 900	1,800	2,700	3,600	4,500	5,400	6,300	7,200 8,100

In this musical interval there are relatively few overtones common to both tones. As a result this musical interval sounds dissonant. For musical intervals having larger whole number ratios we find fewer and fewer overtones common to the tones being played together.

The Musical Fifth Matches the Five Tanmantras

At this stage, we can draw our first major correspondence between the dynamics of music and the primal pattern responsible for the genesis of the material universe. According to the Vedic tradition, consciousness, or the unified field of intelligence, lies at the basis of creation. The five fundamental categories of matter and energy called the tanmantras emerge from this field. A similar process is also found in the construction of the basic musical intervals. As already stated, a musical tone is the fundamental building block of music. However, musical tones by themselves do not create music; it is the relationship between musical tones that gives rise to the sensation of music.

We propose that the key musical interval to emerge from the relationship of musical tones is the fifth for three reasons. First, although the unison and octave are considered to be more basic intervals than the fifth, they serve a more "structural" role in music. The unison is essentially a single musical tone, and the octave provides the framework on which the order of music revolves. Neither the unison nor the octave alone are able to construct the other intervals, scales or chords of music; however, the musical fifth can be used to construct other intervals, scales and chords. Hence, the properties of the fifth match the nature of the five tanmantras which are to be active processes that create, construct and build. (The importance of the octave was illuminated in Chapter 5 of this book with relation to the Constitution of the Universe.) Second, the musical fifth is described by Helmholtz as a "perfect consonant" and is the smallest whole-number interval following the unison and octave. Musical intervals of whole number ratios larger than the fifth become increasingly more dissonant. Third, the fifth is defined as the first and fifth notes of a scale sounded together. For most scales this means that there are exactly five notes that make up the fifth interval. Quantitatively this exactly matches the five tanmantras.

Figure 44 summarizes this parallel between the five tanmantras and the five notes which exist in the musical fifth (from C to G) on the Just or Natural Scale. The ratio of that particular tone with reference to the first tone of the scale, "C," is indicated above each note. We have also given the musical ratio or interval for each pair of notes in this sequence. (Note: the five notes of the fifth interval are written in descending order for reasons that will become evident shortly. Whether the notes of a scale are written in ascending or descending order is a matter of convention.)

Construction of the Major Triad Corresponds to the Creation of the Three Doshas

So far we have considered the effect of sounding together two musical tones. Now, it is reasonable to examine briefly what happens when more than two tones are sounded together. The sounding of a group of musical tones is called a chord. As with the sounding of two musical tones, chords are classified as consonant or dissonant. Not surprisingly, one of the most basic and pleasing chords is the major triad, a three-note combination. The major triad has been the foundation of Western music for several hundred years (Backus, 1969; Hindemith, 1945) and consists of first, third and fifth tones of a scale. According to Helmholtz, the major triad is distinguished from all other triads because the tones have the smallest intervals with one another and is thus considered to be the fundamental chord or basis of all other major chords (Helmholtz, 1954, p. 212).

Although a major triad can be constructed with any three notes whose frequencies are in the ratio of 4:5:6, we generally write the major triad as consisting of the notes G - E - C. The interval between the first and second tone is 4:5 (a major third), and between the first and third is 2:3 (a fifth). Thus, to build a simple triad we can take any note, for example C, and add two tones above it at intervals of a third and a fifth. The musical fifth in a triad is always perfect, and the third can be either major or minor. Whether the third is major or minor will determine whether the chord is a major or minor triad.

Using the major triad as a basis, we can construct more complex chords, scales and other musical combinations. It should be noted for the technical reader that because the third is not properly in tune in the Pythagorean Scale, a triad in this system sounds discordant. However, in the Just or Natural Scale the triad is in tune and produces a very beautiful, fulfilling and consonant chord. For this reason we have used the Just musical intervals in this chapter.

Based on this overview of the fundamental dynamics of music theory, we can draw our second and final correspondence with the Vedic understanding of the mechanics responsible for the genesis of the material universe. According to Vedic understanding, after the five tanmantras have emerged from the unified field of intelligence, they merge together in a well-defined pattern to create three new entities called doshas. The way they merge is as follows: akasha and vayu tanmantras merge to form vata dosha; agni and jala tanmantras merge to create pitta dosha; and jala and prithivi tanmantras merge to produce kapha dosha. Similar dynamics are also found in music theory. Once we have established the musical intervals that can be constructed when two tones are played simultaneously, we may then examine what happens when more than two tones are sounded together.

Since it is possible to construct a major triad from a fifth (and a third) we propose that these two fundamental elements of music (the major triad and the fifth) match the three doshas emerging from the five tanmantras according to the Vedic tradition. In Figure 44 the five notes of the fifth interval and the three notes of the major triad are arranged in such a way that the five tanmantras merge to form the three doshas. Although the correspondence between the five notes of the fifth merging to the three notes of the major triad is not exact, there are a number of points that support it. First, it is possible to construct the major triad using the fifth and a third. The interval between the first note and the third note of the major triad is 2:3, the same as the fifth interval. Hence the two tones of a fifth (C and G) are also the first and third notes of the major triad. The middle note of the major triad (E) is also the middle note of the tones lying in the fifth interval. Second, the five notes found in the fifth interval appear to form two groups based on the greater consonance. The first group, the notes G and F, have an interval ratio of 9:8. The second group consisting of the notes E and D, and D and C have interval ratios of 10:9 and 9:8, respectively. All these are relatively small number ratio intervals (and thus more consonant) compared with the interval between F and E which is 16:15. These two groupings of notes in the fifth interval suggest the pattern observed when the five tanmantras merge to the three doshas. However, the pattern is not exact. Questions still remain why the tone F would combine with G to give G, why the tone D would combine with the tone E to give E and why also with C to give C.

In summary, we have discussed the basic laws of nature governing the sensation of sound in music theory. These principles are universal and independent of the musician or culture in which the music is played. Further, we presented evidence that there is a fairly good match between the construction of the fundamental intervals and chords of music with the structure responsible for the creation of the material universe as understood by the ancient Vedic civilization. Using these basic tonal relationships, musicians create the diverse aesthetics of melodies, rhythms, tunes in the world of music.

CHAPTER 15:

ART: THE FIVE FUNDAMENTAL COLORS MERGING TO FORM THE THREE PRIMARY COLORS

Although the use of colors has always been a major avenue of creative expression, artists have generally avoided understanding color in a scientific manner. Instead, artists have relied heavily on intuitive abilities to understand how to use colors to create harmony and beauty. Consequently, much of our current knowledge of color theory is attributed to the work of scientists. Since the initial work of Sir Isaac Newton, there has been substantial progress in color theory resulting in three major systems: the color system of Wilhelm Ostwald, the color system of Albert Munsell, and the red-blue-green primary color theory. Earlier in this book we discussed the basic principles of Ostwald's color system and demonstrated that they matched the numerical structure of the Constitution of the Universe template. In this chapter we build on the information discussed in the earlier chapter and reveal yet another correspondence between the color systems of art and ancient Vedic wisdom. This new correspondence uses knowledge from the other two color systems, Albert Munsell and the three primary colors theory, to demonstrate how the dynamics of color closely reflect the numerical pattern responsible for the genesis of the material universe.

The Munsell Color Solid

Albert Munsell is considered one of the greatest American color theorists for his system of naming colors which has universal appeal for both layman and specialist. His system of color notation, developed around 1900, is the most widely known and lends itself well as a research tool by scientists. It is used extensively throughout the world and has been adopted by the American Standards Association as industry's language and standard of color. One of the key values in Munsell's color notation system is that it names colors in a logical and orderly way. For example, prior to Munsell, colors were given names such as "forest green" or "cherry red." Although these names are very descriptive, they have little value in specifying or matching colors precisely (Birren, 1969b).

Like Wilhelm Ostwald and other color theorists, Munsell saw the world of colors arranged in three dimensions as a color solid. Building on the work of other theorists such as Hermann Helmholtz, Munsell classified colors in his color solid according to three dimensions of quality: hue, value and chroma (Birren, 1969b). From these three characteristics all colors can be derived and distinguished from one another. The first characteristic used to describe a color, called the "hue," is the quality that distinguishes one color from another (i.e., the name of the color). For example, a color may be yellow, blue or green, etc. The second way to describe a color, called the "value" or lightness of a color, allows us to distinguish a light color from a dark color. For example, a light red color is referred to as pink and a dark red may be called maroon. The last way to describe a color, called the "chroma" or intensity of a color, depicts the amount of gray in the color which determines whether the color is dull or vivid.

The best way to conceptualize a Munsell color solid is to think of it as a sphere-like shape in which the world of colors can be located. The central axis of this color solid is called the "gray scale." This scale consists of ten achromatic colors (based on the decimal system). The end-points consist of "perfect" black at zero (bottom of the axis) and "perfect" white at ten (top of the axis). We say "perfect" white and black because such colors are not possible. Between these two extremes are the gradations from white to black, with neutral gray in the middle at color number five.

According to Munsell, the color neutral gray was considered the most important factor in color harmony and balance. Gray was used to balance colors that are either too strong or too weak so that they become pleasant to look at. For Munsell, a student should first master the simplest and most elementary form of balance, neutral gray, before proceeding to investigate more interesting centers of balance. For instance, one fascinating point is that, when complementary colors are mixed together, they form neutral gray (refer to Chapter 6, Section I).

Projecting outwards from neutral gray (the middle of the axis of the color solid) is a circular plane. To conceptualize this, think of the Munsell color solid as a globe. The gray scale forms the axis with the color white at the North Pole and the color black at the South Pole. Through this axis is a circular plane whose perimeter would correspond to the globe's equator. On this plane Munsell placed what he considered the five fundamental colors: yellow, green, blue, purple and red (Nickerson, 1946). These five colors are arranged in a circular sequence on the plane and represent the hue dimension of color solid. Interestingly, when these five colors are spun on disks they form the color sensation of neutral gray, the color which is located at the center of the Munsell color solid (Birren, 1963).

We have now constructed a basic framework consisting of the gray scale and the five key hues. From these basic components we can proceed systematically to derive all other colors. To begin we can construct a gradation of new colors between each of the five fundamental hues and the gray axis. Each of these new series of colors would start with the pure hue and each new color would be increasingly more dull (increased content of gray) until we reached the neutral gray of the central axis. These gradations of colors represent the chroma dimension of the color solid.

In the next stage we construct gradations of new colors from all colors in the hue and chroma dimensions. Each color in these two dimensions can be further elaborated in the vertical direction (i.e., towards the white and black of the gray scale). For each of these new series the colors would become progressively lighter as we proceeded to move "upwards" until we reached white. Similarly, in each new series the colors would become increasingly darker as we proceeded "downwards" until we reached black. These new gradations or series of colors represent the Value or Intensity dimension of the Munsell color solid. A very practical system of color notation arises from this Munsell color solid. Each new color is written according to its hue, chroma and value content.

For the technical reader we will add one additional point of detail. Different hues have different chroma strengths. For example, the strongest red is twice as powerful as the strongest blue-green pigment. Consequently, the number of gradations of colors arising from one hue will differ from another hue. This gives the Munsell color solid the appearance of a "color tree" with the trunk representing the gray scale and each branch a different hue. The branches vary in length according to the chroma strength of each hue (Birren, 1969b).

The Color Sensation (Samhita) is Described by Hue, Value and Chroma

Now we can draw our first set of major correspondences between the basic principles of Munsell's color system and the numerical structure that gives rise to the material universe. Figure 46 presents a summary of all these correspondences. Before we begin however, we need to point out that in this chapter we are considering *colored light* rather than *color pigments*; this is for several reasons. First, the perception of color is a highly personal experience – what one person calls a particular color may not be the same as what another person calls that color. In other words, color is a sensation and is a personal experience of radiant energy. Every color we see enlivens a different emotion and aspect of our consciousness. The color sensation is a result of seeing colored light. We do not see the actual color of the pigment, but rather the light that is reflected from it. In fact, light is the only thing we can ever see in the world around us. But like other words in the English language, light has a variety of meanings. For example, physicists will treat it scientifically, religious thinkers will view it symbolically and artists will use it practically. Another reason we choose to use colored light in this section will become evident when we explain the result of mixing colored light. Mixing together pigments instead would give different outcomes. In relation to the Vedic understanding of the genesis of the material universe, color sensation corresponds to the samhita value of knowledge representing the summation of rishi, devata and chhandas.

As discussed earlier, all colors are described in three distinct ways which prompted us to associate them with the flavors of rishi (knower), devata (process of knowing) and chhandas (known). The first characteristic used to describe a color, called the "hue" of the color, corresponds to the nature of rishi which is equated with the inherent or abstract nature of an entity. The second way to describe a color, called the "value" or lightness of a color, is more the nature of devata. The last way to describe a color, called the "chroma" or intensity of a color, matches the nature of chhandas.

Munsell's Five Basic Colors Match the Five Tanmantras

Above "hue," "value" and "chroma" we have placed white light (refer to Figure 46). In the work of Munsell, and other leading color theorists,

Figure 46. Genesis of the World of Colors According
to Art: The Three Primary Colors Arising
from the Five Fundamental Colors

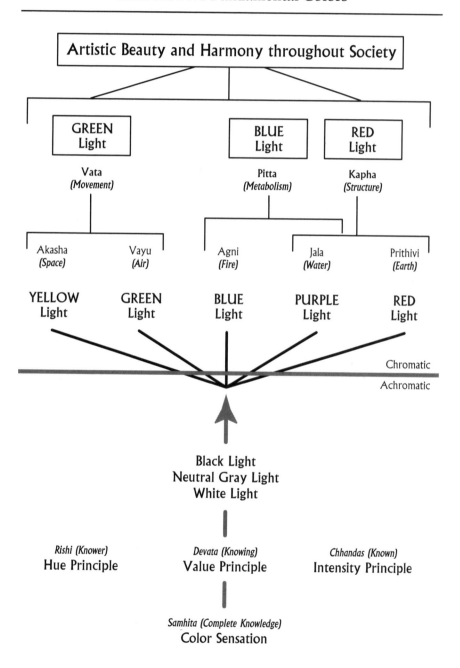

white is located at the top of their color solids. From the angle of physics white light contains all other colored lights. It is also interesting to remember that in the Holy Bible the following is stated with reference to the first day of Genesis, "And God said, let there be Light. And there was Light." Perhaps pure white light stands at the doorway of the Absolute.

The gradation of achromatic colors is constructed from white light. In this series of colors we have neutral gray and finally the color black. This "gray scale" is represented at the bottom of Figure 46. It is important to realize that white and black are colors, as artists have unanimously insisted. The sensation we get from looking at a black color, for example, is quite different from the absence of all sensation.

Since Munsell's color solid is based on five key hues, we propose that these five basic colors match the five tanmantras or fundamental categories of energy and matter from Vedic knowledge (refer to Figure 46). In addition to the exact quantitative correspondence between Munsell's five key hues and the five tanmantras, there are also a number of qualitative reasons giving support to this correspondence. First, spinning the five key hues on disks gives the sensation of neutral gray, the color located in the center of the Munsell color solid. Neutral gray is also considered the color of balance and is formed when two complementary colors are mixed together. Second, the five Munsell hues turn out to be exactly the major color constituents of solar light which the human eye can discriminate (Life Magazine, 1944). Third, Munsell's color system is internationally accepted, and its nomenclature has been adopted by many organizations including the United States Bureau of Standards.

Formation of the Three Primary Colors Corresponds to the Three Doshas

It is possible to derive the three primary colors of light (red, green and blue) from Munsell's five key hues. Many scientists, including Munsell (to some extent), have felt that the "red-blue-green theory" of color is fundamental. These three colors are all that are needed to derive the world of colored lights. Also, these three colors form the basis for color photography and color television.

The construction of the three primary colors of light from Munsell's five key hues closely matches the creation of the three doshas from the five tanmantras as described in the ancient Vedic Literature (refer to Figure 46). We begin our comparison with the construction of the primary colors blue light and red light from Munsell's blue, purple and red. When blue light and red light combine in a 50:50 mixture they create the color purple. Thus, the primary color red can be derived from Munsell's pure red and the 50% red component of Munsell's purple. Similarly, the primary color blue is derived from Munsell's pure blue and the 50% blue component of Munsell's purple. The last primary color green light is derived from the remaining two Munsell colors (green and yellow). Yellow light is the result of green light and red light combining in a 50:50 mixture. Hence, the primary color green can be derived from Munsell's pure green and the 50% green component of Munsell's yellow. (We do realize that the other 50% of Munsell's yellow is the color red which would contribute to the primary color red. This fact suggests the possibility of a link between the Munsell yellow and red, and the primary colors green and red. Both sets of colors could thus be arranged into hue circles instead of linear series. The idea to arrange hues in a circle is common to most color theorists.) Another point supporting the construction of the primary color green is that the color yellow has a tendency to be quickly overshadowed by the addition of even small amounts of other hues. Together the three primary colors when spun on disks create the sensation of white light.

The construction of these three primary colors of light match the well-defined pattern of the five tanmantras merging to create the three doshas according to the Vedic tradition. This merging is as follows: Akasha and Vayu tanmantras merge to form vata dosha; Agni and Jala tanmantras merge to produce pitta dosha; and Jala and Prithivi tanmantras merge to create kapha dosha. Apart from the quantitative aspect of this final correspondence between the dynamics of color and mechanics responsible for the genesis of the material universe, there is also one qualitative parallel worth mentioning. In Ayurveda (the ancient medical system of Vedic India) the three primary colors are associated with the three doshas (e.g., in color therapy) in the exact same sequence as we have presented them in Figure 46.

In conclusion, we have cited a clear correspondence between the dynamics of colors and the numerical pattern responsible for the genesis of the material universe known to the ancient Vedic civilization. Through the proper dynamics of color, art is created which is balancing, harmonious and uplifting for all levels of society.

CHAPTER 16:

ANCIENT GREECE: THE FIVE PLATONIC SOLIDS MERGING TO FORM THREE BASIC SHAPES

"When God called the world into existence, he worked as a mathematician."

– Pythagoras

"Geometry existed before the Creation."

– Plato

According to the ancient Greeks, numbers are the secret and origin of all things. They believed that the different forms of numbers were the key to understanding the spiritual and physical universe. The ultimate reality is not material, but spiritual; it consists of the concepts of numbers and form. These ideas were clearly expressed by Plato who believed that the universe was intelligible. In Timaeus, his dialogue on natural science and cosmology, Plato described five basic geometrical solids which account for the creation of the universe. These five solids (called the "Platonic Solids" in honor of Plato), the core and foundation of Sacred Geometry, recreate primal order out of chaos. In this section, we will discuss the central ideas underlying the five Platonic Solids and demonstrate that the dynamics of these five solids seem to match the numerical pattern responsible for the genesis of the material universe.

The Sphere (Samhita) Contains the Image of All Other Forms

Figure 47 presents an overview of the correspondences between the emergence of the five Platonic Solids and the structural mechanics responsible

Figure 47. Genesis of Creation According
to the Ancient Greeks: The Three Primary
Shapes Arising from the Five Platonic Solids

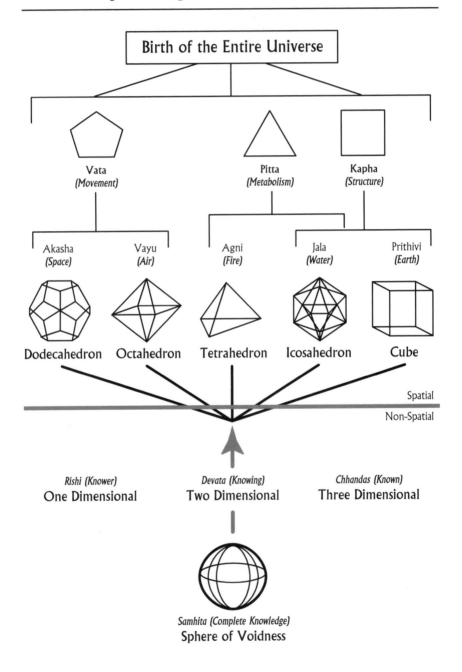

for the genesis of the material universe to be discussed in this chapter. The purest and most ideal form – the sphere, is at the bottom of this figure. According to Plato, the sphere is the first shape which God created and in its image all other forms are comprehended. Since the sphere represents the summation of all things, it is the key to wisdom and the "God of all gods" (Lawlor, 1982). An interesting component of a spherical body is that all points on its surface are equidistant from the center. In relation to the Vedic understanding of the genesis of the material universe, the sphere corresponds to the samhita value of knowledge representing the together-ness of rishi, devata and chhandas.

In the construction of a sphere we automatically introduce the concept of three-dimensional space. The three dimensions correspond to the three aspects of rishi (knower), devata (process of knowing) and chhandas (known). The first dimension is the most subtle and is represented by a straight line which corresponds to the abstract nature of rishi. The second dimension, representing the concept of area, acts as a transitional dimen-sion between the first and the third dimension. This corresponds to the qualities attributed to devata which are to connect or act as a transitional state. Last, the third dimension represents the most concrete expression of the three dimensions and is thus aptly equated with the material nature of the chhandas interpretation.

The Five Platonic Solids Correspond to the Five Tanmantras

Out of the sphere, representing a state of voidness, there are created five fundamental forms responsible for the structure of the universe. Each form possesses volume, and every volume is bounded by surfaces. The geometrical dimensions of these forms are the five Platonic Solids. How these five solids can be constructed from a sphere was outlined earlier in this book (refer to Chapter 8, Section V). In that section we described how to construct the Flower of Life diagram which provided a framework for creating the five Platonic Solids.

The five Platonic Solids include: dodecahedron, octahedron, tetrahedron, icosahedron and the cube (refer to Figure 47). Each of these solids satis-fies four conditions or constraints: first, all edges on the solid are equal; second, all angles on the solid are the same; third, all the faces on the solid are identical; and fourth, when placed in a sphere all the vertices of the solid must touch the surface of the sphere. In fact, the five Platonic Solids are the only polygons which meet all four of these constraints. From these five solids all other regular polyhedra can be constructed.

The following tabulates the number of vertexes, faces and edges for each Platonic Solid:

	Number of Vertexes	Number of Faces	Number of Edges
Dodecahedron	20	12	30
Octahedron	6	8	12
Tetrahedron	4	4	6
Icosahedron	12	20	30
Cube	8	6	12

At this stage we can draw our first major correspondence between the dynamics of the Platonic Solids and the structural template responsible for the genesis of the material universe. According to Vedic wisdom there are five fundamental categories of energy and matter, called tanmantras, which emerge from a unified field of intelligence. Similarly, the ancient Greeks claimed that there were five fundamental solids which emerged from the sphere of voidness. From these five Platonic Solids the universe was created and is maintained. Apart from the exact numerical correspondence between the five Platonic Solids and the five tanmantras, there is also an exact qualitative correspondence. Plato describes this in Timaeus where he states that each of the five universal elements are expressed or find their form in each of the five Platonic Solids (Guthrie, 1988; Heninger, 1974; and, Lawlor, 1982). The assignment of the five Platonic Solids to the five elements was not an arbitrary decision, but was based on the following rationale:

- The cube or hexahedron was assigned to the Earth element because it is dense, flat and gives the impression of stability which is indicative of terrestrial matter.

- The icosahedron was assigned to the Water element. Of all five Platonic Solids the icosahedron has the largest number of faces, suggesting "a water-drop" (Heninger, 1974).

- The tetrahedron was assigned to the Fire element. Interestingly, the word for pyramid means "fire in its midst." This geometrical solid is considered to be the primal shape from which the entire universe can be constructed. Similar ideas have also been advanced by Buckminister Fuller who believed that the tetrahedron was the most fundamental unit of creation (Fuller, 1975).

- The octahedron was assigned to the Air element. There are several reasons for this. First, the "octahedron can be suspended by two opposite corners and spun as in a lathe, thereby representing 'a certain image of mobility' suitable to air, the most mobile element" (Heninger, 1972, p. 107). Second, the eight edges of the octahedron may relate to oxygen, a component of air and also the eighth element on the periodic table of elements (Melchizedek, 1992).

- The dodecahedron was assigned to the prana or ether element. This solid was considered the most sacred shape in the cosmos and was said to contain all other shapes. It was a symbol for the universe because its twelve faces were believed to correspond to the twelve constellations or twelve signs of the Zodiac.

Hence, it is clear that the five universal elements that Plato associated with his five geometrical solids exactly match the translated names of the five tanmantras. We feel that this gives substantial credence to the importance of Plato's wisdom as distinct from our modern views. Typically, modern scientists have been quite naive in their views of ancient Greek knowledge and have often considered the five universal elements as a "primitive" periodic table of elements. This view is not only supercilious, but vastly underestimates what ancient civilizations understood about the origin and dynamics of our universe.

Three Polygons are used to Construct the Five Platonic Solids

At first the five Platonic Solids appear unrelated to one another, except by the fact they are geometrical solids and are composed of regular polygons. However, in Timaeus Plato describes in detail the intermingling of the five solids and five elements. Although his writings describe interactions occurring between all the elements, the following stand out as being clear illustrations. First, the relationship between the elements water and earth. When the element water condenses it forms ice, snow or hail which exhibit the solid properties of the element earth. Also, water is said to be the only element that is able to "melt the masses of earth." The movement of water is capable of forcing a passage into the ground and dissolving away the earth. Second, the relationship between the elements water and fire. Movement of water along the ground resembles the dynamic flames of the fire element. In addition, there are some types of water mixtures such as wine which have a fiery nature. Third, the relationship between the elements air and ether. As with the other elements, there are different types of air. One of these types is located in the upper portion of our atmosphere and exhibits many of the qualities of outer space or "ether." As already mentioned, these three examples are not the only relationships among the elements, but they do stand out as clear illustrations.

These relationships between the five Platonic Solids and the five elements prompted us to look for some way that they could merge to form three basic entities as in the Vedic understanding that the five tanmantras merge to create the three doshas. The five tanmantras merge in the following way: Akasha and Vayu tanmantras merge to create vata dosha; Agni and Jala tanmantras merge to create pitta dosha; and Jala and Prithivi tanmantras merge to produce kapha dosha. As discussed above, it is possible to find three clear descriptions from Plato's writings of how his five elements intermingle or merge in an equivalent pattern as the five tanmantras to the three doshas. However, a number of additional relationships between the elements are possible which do not lend support to the exact order of merging from the Vedic understanding.

The answer to the perplexing question of how the five Platonic Solids merge to form three basic entities is found in the geometry of these shapes. In fact, only three polygons are needed to construct the five Platonic Solids – the triangle, the square and the pentagon. The dodecahedron is constructed of pentagons, the cube of squares, and the icosahedron, tetrahedron and octahedron of triangles. Our proposed arrangement of these three fundamental shapes as they relate to the five Platonic Solids is summarized in Figure 47. This arrangement is based on the Vedic understanding of how the five tanmantras merge to form the doshas. Although this correspondence is not completely precise between the five Platonic Solids and three basic polygons (triangle, square and pentagon), there are a number of arguments that support it. First, the dodecahedron is an obvious candidate for merging to the pentagon shape. Second, the cube is clearly associated with the square shape. Third, the faces of the tetrahedron and icosahedron are very well associated with the triangle shape. The remaining correspondences remain obscure – how does the octahedron relate to the pentagon shape, and how does the icosahedron relate to the square? Although a substantial volume of literature is published on the geometrical interactions and relationships between the five Platonic Solids, no clear answer to these two questions is revealed. However, we feel that there probably is some deep significance that has been lost over time. Perhaps the answer to such questions will help reveal additional insights into the significance and dynamics of the Platonic Solids.

In conclusion, we have presented an interesting correspondence between the origin and dynamics of the five Platonic Solids known to the ancient Greeks and the pattern responsible for the genesis of the material universe known to the ancient Vedic civilization. This correspondence may help to elucidate further the importance of these five geometrical solids which have been a long-time favorite for those studying Sacred Geometry.

CHAPTER 17:

ANCIENT CHINA: THE FIVE ELEMENTS MERGING TO FORM THE THREE BASIC AXES

The ancient Chinese understood that change was a constant phenomenon; it was always happening. Out of this seeming chaos of the universe, however, they observed cycles and fundamental principles of order governing the destiny of all events. In particular, they classified all processes and all substances according to five fundamental elements or interacting forces of nature. These five basic elements can be seen at work in everything and combine together in innumerable ways to create the myriad of diversity. In this chapter we will describe in detail these five fundamental processes permeating the whole of Chinese culture. Further, we will demonstrate that the dynamics and origin of these five elements match exactly the structure responsible for the genesis of the material universe as understood by ancient Vedic wisdom.

The "Three Pure Ones" are Created from Wu Chi and the Tao (Samhita)

Figure 48 presents a summary of all the correspondences between the dynamics of the five interacting forces of nature known to the ancient Chinese and the structural pattern responsible for the genesis of the material universe. At the bottom of this figure we have what the ancient Chinese considered to be the ultimate reality, "Wu Chi." This was regarded as the basis of all existence, unifying all objects and phenomena. Symbolized by an empty circle, Wu Chi represents the unmanifest which is beyond the manifest. Although this level is beyond description in words, it is often known by the following terms: the supreme ultimate, the original emptiness, the purest original force or the nameless one (Chia and Chia, 1989).

Figure 48. Genesis of the Universe According to the Ancient Chinese: The Three Basic Axes Arising from the Five Elements

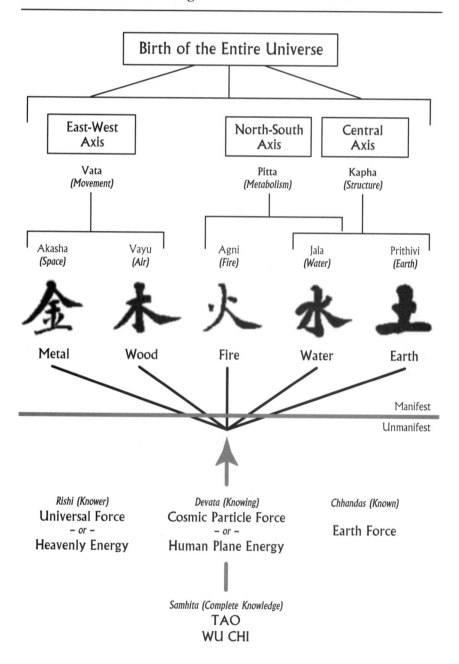

Wu Chi creates the universe through the primal mechanics known as the Tao. The Tao (pronounced as "dow") is translated as meaning the "Way" and is considered to represent the dynamic path of unfolding or internal dynamics inherent in Wu Chi (personal communication, Douglass White). In the words of the famous Lu Tzu, "That which exists through itself is called the Way (Tao). Tao has neither name nor shape. It is the one essence, the one primal spirit. Essence and life cannot be seen." (Wilhelm and Jung, 1962). Fundamentally the Tao, although itself motionless, is the source of all movements and ways of nature. In relation to the Vedic understanding of the genesis of the material universe, Wu Chi and the Tao correspond to the samhita value of knowledge which represents the summation of rishi, devata and chhandas.

Out of the Great Emptiness of Wu Chi and the internal dynamics of the Tao, the first energies to emerge are called the "Three Pure Ones" (Chia and Chia, 1989). These three fundamental energies correspond precisely with the flavors of rishi (knower), devata (process of knowing) and chhandas (known). The first of the Three Pure Ones is called the Universal or Original Force, also known as the Heavenly Energy. It manifests as the energy of galaxies, solar systems and stars acting to nourish the mind and spirit of everything in the cosmos. This force matches the nature of rishi which is associated with knower (mind) or abstract energy. The second of the Three Pure Ones called the Cosmic Particle Force or the Human Plane Energy, is believed to be the force behind exploded stars. The fine dust particles left behind by such stars create the energies of the human body. Human energy connects the heavenly energies (first Pure One) with the earthly energies (third Pure One). This matches well with the nature of devata which is to link rishi and chhandas. The third of the Three Pure Ones is known as the Earth Force which is responsible for the energy of all things which occur on Earth. This corresponds to the material or earthly qualities of chhandas. Together the three forces represent the energies of heaven, humans and earth working in harmony to sustain all existence (Chia and Chia, 1989).

Description of the Five Elements and Their Correspondence to the Five Tanmantras

From the triple unity of the Three Pure Ones the five elements are created (refer to Figure 48). The ancient Chinese believed that "... the Five

Elements of Nature originated as five huge stars, given birth by the Three Pure Ones out of the Wu Chi. These five stars (or five elements), in turn, gave birth to the entire universe, including trillions of stars." (Chia and Chia, 1989 p. 5) All of the five elements are understood to be basic dynamic processes rather than passive objects. These five elements are as follows: Metal, Wood, Fire, Water and Earth (the Chinese symbols for these elements are displayed in Figure 48). It should be noted, however, that English names for these five elements should not be taken as literal translations since the elements represent dynamic states of matter. A brief description of each of the five elements follows (Connelly, 1979):

- The element Metal is often the most difficult to experience in Nature. Typically this element gives the impression of cold, hard and non-living. There are, however, a number of other qualities attributed to this element: it represents strength, force, and structure.

- The qualities of the element Wood are easy to grasp if we think of a tree. As a tree grows it is flexible, strong, durable and yields to the wind. Wood expresses itself in growth and the hope for future harvest. It also embodies the characteristics of solidity, workability and steadfastness.

- The qualities of the element Fire include the following: heat, light, bright, warmth, dynamic, moving, liveliness, joy, love and compassion. This element can be used to create and direct energy.

- The element Water is considered to be the life-giving principle, for without water life as we know it on Earth would not exist. The descriptions of water in nature are the same as the descriptions of water within us – inside our physiologies we have ponds, rivers, seas, oceans of energy and sources of life. It also represents the qualities of force, power and movement.

- The last element, Earth, is responsible for the production of edible and nutritive vegetation. It is "Mother Earth" from which we obtain our nourishment, support and life. Other qualities associated with this element include fertility, stability and fullness.

All these five elements are present in every aspect of the universe. Every part of our physiology is supported by the interaction of these five fundamental forces. The knowledge of which elements reside in which parts of the body forms a substantial part of ancient Chinese medicine. In addition to understanding the association between the human body and the five elements, the Chinese proceeded to classify all known entities into groups of five. With these groupings they hoped to elicit order from the seeming world of chaos. For example, they discovered that there were five basic cardinal directions, five colors, five seasons, five senses, five planets and many other groups of entities which they related to the five fundamental elements. (Nylan, 1994).

At this stage we can draw our first major correspondence between the origin and dynamics of the five elements and the five tanmantras (refer to Figure 48). According to Vedic wisdom there are five fundamental categories of matter and energies, called tanmantras, arising from a unified field of intelligence. Likewise, the ancient Chinese understood that out of the Great Emptiness, Wu Chi, exactly five fundamental processes of the universe emerge. Not only is this an exact quantitative correspondence between the Vedic and Chinese systems, there is also a very close qualitative parallel. Three of the five Chinese elements (Fire, Water and Earth) have the exact same names as the last three tanmantras. The Chinese element Wood is often associated with the image of a tree bending in the wind which corresponds to the second tanmantra of Vayu or Air. The last Chinese element Metal embodies the qualities of cold and non-living. This could be associated with the first tanmantra, Akasha or Space. Space is an arena of coldness, emptiness and, except for the occasional star cluster, devoid of life.

Construction of Three Axes from the Five Elements Matches the Three Doshas

A number of other scholars have already observed the close correspondence just described between the five tanmantras and the five elements (e.g., see Amber and Babey-Brooke, 1966). It appears, however, that no one has been successful in furthering this correspondence to show how the five elements merge to some basic three in the same way that the five tanmantras merge to create the three doshas. After some consideration we discovered a way in which such a correspondence can be achieved.

The answer could very well lie in the five cardinal directions which the ancient Chinese associated with the five elements (personal communication, Douglass White). Figure 49 summarizes how each cardinal direction corresponds to a particular element. These five cardinal directions merge

Figure 49. A Diagrammatic Representation of the Five Cardinal Directions and the Three Fundamental Axes on a Planetary Sphere

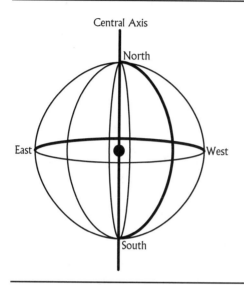

Five Elements and the Cardinal Directions:
* Metal = West
* Fire = South
* Earth = Center
* Wood = East
* Water = North

Three Fundamental Axes:
* East — West
  (longitude lines)
* North — South
  (latitude lines)
* North — Central
  (axis through the center)

to create three new entities through the construction of axes. Prior to performing this operation, however, we need to realize that the ancient Chinese probably used the five cardinal directions with respect to our Earth, a planetary sphere, and not to a two-dimensional plane. From such a model, three unique axes can be constructed (refer to Figure 49).

According to the ancient Chinese, the first axis to be constructed is the East-West axis from the Metal (West) and Wood (East) elements. This axis corresponds to the latitude lines that span the outer surface of the sphere. The second axis to be constructed is the North-South axis from the Fire (South) and Water (North) elements. This axis corresponds to the longitude lines that span the outer surface of the sphere. The last axis is more subtle to construct and is the result of Water (North) and Earth (Center). Many scholars seem to have encountered a problem interpreting the association of the Earth element with the "Center." We think this is because the five cardinal directions have been viewed on a two-dimensional surface. In two dimensions an axis constructed from the Center to the North would be identical to the North-South axis. In a three-dimensional sphere, however, the center-North axis would run through the middle of the sphere, whereas the North-South axis would be the longitude lines on the outer surface. A summary of these axes is presented in Figure 49.

Construction of these three axes from the five elements matches the well-defined pattern in which the five tanmantras merge to create the three doshas according the Vedic tradition. This merging is as follows: akasha and vayu tanmantras merge to form vata dosha; agni and jala tanmantras merge to create pitta dosha; and jala and prithivi tanmantras merge to produce kapha dosha. This same sequence is observed with the five elements of the ancient Chinese merging to create the three axes (refer to Figure 48). It should be mentioned, however, the three basic axes are not generally recognized as serving a central role in Chinese philosophy. Even so, this is an exciting correspondence considering the well-defined pattern by which the five elements merge to create the three axes. Further, the construction of the three axes may shed light on the ancient Chinese assignments between the five elements with the five cardinal directions. Perhaps there is a deeper significance to these axes still to be discovered.

In conclusion, we have presented an intriguing correspondence between the origin and dynamics of the five elements known to the ancient Chinese and the genesis of the material universe known to the ancient Vedic civilization. This provides additional support to the theory that all the ancient civilizations had a similar understanding of the primal mechanics responsible for the universe.

CONCLUSIONS
AND REFLECTIONS

"But seek ye first the kingdom of God, and his righteousness; and all these things shall be added unto you."

Holy Bible, Matthew 6:36
(Authorized King James Version)

It is evident from the foregoing chapters that the concept of the Constitution of the Universe is both all-embracing in its application and ageless in its wisdom. We see the structure of the Constitution of the Universe displayed everywhere simply because there is order and intelligence throughout creation. As we have demonstrated, there are numerous ways in which the Constitution of the Universe is expressed, in all different languages, symbols, disciplines, religions and terminologies. In each of these illustrations we have found that the most fundamental aspects and entities of life have common ground in the structure of the Constitution of the Universe. Arising from these primal levels, all diversity can be constructed, organized and governed. Based on this research, there is enough evidence to establish the authenticity and complete understanding of the Constitution of the Universe as nature's most elementary pattern of intelligence that determines all entities and explains the origin of creation. We suspect that this clear and simple comprehension of the foundations of universal wisdom will provide an avenue through which all other aspects of the advanced, but lost, knowledge of the ancient world can be restored to their complete values.

Intellectual and Practical Applications

From an intellectual standpoint, the Constitution of the Universe provides a powerful and unifying educational tool. By demonstrating the recurrence of the structure of the Constitution of the Universe throughout the fundamental aspects of the academic disciplines we can readily see the seemingly dissimilar fields of learning are in fact, all interconnected. We find, therefore at the foundation of all disciplines and ancient teachings there is a common language linking everything together. Thus, there must be a unified level of nature's dynamics from which all diversity is created and governed. According to ancient wisdom, this unified level of life was believed to be a universal field of consciousness and intelligence. Through advanced spiritual programs the Ancients were able to explore this field of consciousness and obtain profound insights into how creation worked. It is the wisdom from these ancient insights and cognitions that has allowed us now to restore a complete intellectual understanding of the Constitution of the Universe.

There are a variety of practical applications that can be derived from the intellectual understanding of the Constitution of the Universe provided in this book. First, we can use this foundational wisdom to restore other aspects of ancient wisdom and develop the details of how all diversity is created. For example, in the second part of this book we provide a fine illustration of how to derive the primal mechanics for the genesis of the material universe from the Constitution of the Universe. Second, understanding the structure of the Constitution of the Universe provides a tool to make new discoveries and predictions in modern science. Third, using the numerical pattern of the Constitution of the Universe, it is possible to create new links between the various academic disciplines (e.g., developing rules for transforming the gene sequence of DNA into musical scores). Fourth, the structure of the Constitution of the Universe could be incorporated for improved marketing strategies. For instance, many companies are already using the numbers of the Constitution of the Universe to design board games and to package products.

Another significant conclusion from this book is that ancient cultures possessed a profound understanding of the universe. Regardless of how or why the Ancients had such wisdom, clearly it is the most important aspect that has survived from these civilizations. Although the spiritual understanding of life has been almost lost, there is still much remaining that could be of tremendous benefit to our world. This does not mean, however, that the modern day remnants from these ancient cultures should be considered role models since the current understanding of ancient wisdom is fragmented to the point of being useless. It is also not fair to judge the wisdom of the ancient world by the current state of the countries where these cultures were located. By restoring a complete understanding of universal truths we can integrate these values into our lives and into our world. Only by reviving this knowledge can we hope to live the lives of our spiritual forefathers.

An Experiential Appreciation for the Constitution of the Universe

Presenting an intellectual understanding of the most fundamental mechanics of creation was the major purpose of this book. This, however, is only half the story. Complete knowledge of life consists of both intellectual and experiential understanding. For instance, learning in school about our nation's constitution makes it easier for us to be law-abiding citizens. Comprehending laws and regulations allows us to make better use of ourselves and to be of greater assets to our country. Yet, there is something more profound than simply having an intellectual understanding of our nation's constitution. To be an ideal citizen of a country, we need to have the same broad vision and intelligence as the founders of our constitution. Similarly, an intellectual understanding of the Constitution of the Universe will enable us to act more in accord with the laws of nature. All entities in the universe are constantly under the jurisdiction of cosmic laws which govern the trends in everyone's life. Living life in accord with these laws brings about eternal happiness, whereas violating such laws causes suffering. There are many times, however, where just the intellectual understanding of cosmic laws is not sufficient to judge whether our every thought and action will produce life-supporting effects. To be an enlightened teacher or an ideal citizen of our universe we need to have the spiritual experience of Ultimate Reality underlying the structure of the Constitution of the Universe. Developing our consciousness to experience the Ultimate Reality is the only way that our lives will be fully in tune with the divine plan and our actions will be most effective.

In our society today the majority of individuals have little self knowledge. People often live in the presence of themselves throughout their entire life without ever knowing anything about who they really are. When we think about this deeply, all knowledge is with reference to the Self. If we do not know ourselves, then we really do not know anything. For that reason, it has been hard for many seekers to find their place and purpose in our universe amidst the turmoil and complexities of the modern world. As an advanced cultural group we have tried, endlessly in vain, to gain fulfillment from the material world and our machine-driven civilization. However, in the grand scheme of things it does not matter how much fame, wealth or talent we possess. What really matters is finding the true happiness that we all seek. Finding spiritual happiness and the development of consciousness is not something supernormal. We are all born with the innate ability and the apparatus of a human nervous system to experience higher levels of spiritual fulfillment. We do not need to be some genius, ascetic monk or unusual person to have these experiences.

Transcendental Experiences Provide the Path for the Development of Consciousness

Throughout the Far Eastern traditions of meditation and philosophy there are numerous descriptions of deep structures of consciousness and exalted inner experiences. Such experiences were considered to be the most significant events in life and a source of great inspiration. Reports of these experiences have also been recorded in recent history by individuals such as Plato, Maslow, Einstein, Wordsworth and many others. Apart from these individuals, the occurrence of spiritual experiences in our modern world are either quite rare or are not disclosed for fear of disbelief or ridicule. Although many of us may have glimpsed elevated states of divine awareness during our quieter moments, it is clear that to achieve permanent experience of these levels requires the guidance of an enlightened teacher. The key to achieving enlightenment is to practice techniques such as deep meditation that give repeated and regular experiences of higher stages of human development. Such techniques cannot be learned from a book, they have to be imparted from a teacher. To be guided by an enlightened master is the answer to the secret teaching of the ages. Thus, our purpose in this book is not to provide instructions on how to have divine experiences (which requires a spiritual teacher), but rather to provide the detailed intellectual understanding of the most primal levels of creation from which these experiences arise.

The inner experience of different levels of consciousness is not some metaphysical or philosophical invention. These experiences are real and can be had by any individual during any time or at any particular place. In everyday life we are typically aware of at least three states of consciousness: waking, dreaming and sleeping. Beyond these common states there is a deeper field of consciousness which is out of range of our senses and which cannot be accessed through thought, speculation or imagination. Experiencing this deeper field of consciousness is commonly known as transcending (going beyond) all relative experiences. Transcending refers to the shifting of consciousness or awareness away from the ordinary objects of perception towards the source of thought. When we transcend all relative boundaries, we reach a level of awareness where consciousness is aware of itself alone. Experiences associated with this level of consciousness have been expressed by numerous individuals in diverse terminologies and languages (e.g., the Transcendent, the inner life, the Ultimate Reality, the Kingdom of Heaven, "that peace which passeth all understanding," the Tao, Ayin Sof, and divine inspiration).

Since the transcendental aspect of life is abstract, one may wonder if anything is being experienced at all. It is an "objectless experience" of a universal field of consciousness which clearly lies outside the realm of ordinary awareness. Spiritual or transcendental experiences of this universal field of consciousness are reported to be highly rewarding with many practical benefits. Such experiences are not only intrinsically fulfilling and blissful, but can also enliven new modes of perception and cognitive functioning. It is believed that regular transcendental experiences, through techniques such as deep meditation, will promote the development of higher levels of consciousness. In these higher levels of consciousness it is possible to use the full range of human potential and every action is said to be spontaneously in accord with the cosmic laws that govern our universe. Throughout the ancient cultures there are numerous descriptions of the exalted inner experiences and cognitions of timeless wisdom resulting from the development of spiritual values. However, without the direct personal experience of the inner levels of life we can gain only a superficial understanding of universal wisdom through analogies and parables. Thus, to have a complete understanding of the Constitution of the Universe we need both the intellectual knowledge and the transcendental experience of the Ultimate Reality. To comprehend nature fully is, in effect, to re-create the universe in our own mind.

In recent years the advancement of global communication has brought the western world in contact with many Eastern systems of philosophy and meditation (e.g., Transcendental Meditation, Zen, Yoga and others). This broad exposure has created an upsurge of interest in the ancient wisdom of life and the possibility of spiritual fulfillment. As a result, we are in the midst of a massive restoration that has the potential to usher in a new age for humanity. We hope that this book has provided a unified perspective to the growing body of universal wisdom during this time of spiritual regeneration. May the fundamental knowledge of life provided in this book enlighten the seekers of truth to higher levels of intellectual and experiential understanding, generation after generation.

FUTURE RESEARCH

"Wherefore, the things of all nations shall be made known; yea, all things shall be made known unto the children of men. There is nothing which is secret save it shall be revealed; there is no work of darkness save it shall be made manifest in the light; and there is nothing which is sealed upon the earth save it shall be loosed. Wherefore, all things which have been revealed unto the children of men shall at that day be revealed ..."

Book of Mormon, 2 Nephi 30:16-18
Translated by Joseph Smith, Jr.

We suggest two primary avenues for future research. First, simply continue to look for other illustrations of the Constitution of the Universe and the primal pattern responsible for the genesis of the material universe. We suspect that there are numerous examples of these fundamental templates in addition to the ones described in this book. There are many specialized disciplines as well as remote cultures of the ancient world that could be investigated. This research alone could involve many researchers over many years. Second, to use the clear understanding of primal patterns of the universe described in this book as a reference to restore ancient teachings that may have been lost or distorted. In the pursuit of this type of research, it is imperative to be open to seemingly trivial things, no matter how simple they may appear. The following is a list of some ideas, whether for pleasure, for profit or for higher knowledge:

The Constitution of the Universe:
Illustrations of the Number Sixty-Four

The Keys of Enoch (Hurtak, 1977)

In this popular book J.J. Hurtak describes sixty-four keys of light which were given to him from the Galactic Archive Center by Enoch and Metatron. These sixty-four keys are meant to coordinate the sixty-four unique areas of scientific knowledge and prepare humanity for the quantum changes that are to take place. All of the sixty-four keys were also classified into three groups: the Father Universes, the Son Universes, and the Shekinah Universes.

Fundamentals of Christianity (Torrey, 1990)

This is a well-known book which consists of a collection of sixty-four articles on the fundamental ideas and concepts of Christianity. We discovered that these sixty-four articles could be easily arranged into eight groups of spiritual significance. The chapters of this book were written by 60 different authors, all but one (born in 1676) of the 19th century.

Kama Sutras of Vatsyayana (Vatsyayana, 1992)

This is an aspect of the Vedic Literature that describes the "science of love." It is a treatise on men and women and their mutual relationships. According to the ancient texts there are sixty-four different practices of sexual intercourse that are grouped according to eight divisions. Each practice and grouping is also interpreted according to three virtues of living. The knowledge of the Kama Sutras appears to have originated from the wisdom of the Rig Veda.

Tantrism (Feuerstein, 1994)

This system appears to have Vedic origins, similar to the Kama Sutras of Vatsyayana, in which love or sexual energy is believed to be embodied by the great Goddess Durga. Emanating from the Goddess Durga are sixty-four aspects or minor deities possessing various qualities and spiritual functions.

The Constitution of the Universe: Illustrations of the Number Eight

Proposed Octaval-Based System of Currency, Mass, Volume and Length (Buckman, 1909)

In this book Buckman suggests replacing the current decimal system by a more practical and natural system based on the number eight (i.e., an "octaval system"). It appears that our system of measurements was originally founded on the number eight and over the lapse of time became distorted and replaced by a decimal notation. The following are some surviving examples of measurements which were originally based on the octaval system:

British system of volume and mass have fairly complete systems of eight already in use:

8 pints = 1 gallon	16 drachms = 1 ounce
8 gallons = 1 bushel (of wheat at 64 lbs)	16 ounces = 1 pound (14 lbs = 1 stone)
8 bushels = one quarter	8 stones = 1 cwt (hundredweight)

British system of length:

8 furlongs = 1 mile	Division of the inch in current use: 1/64, 1/32, 1/16, 1/8, 1/4, 1/2, 1

Monetary systems:

Spanish, "pieces of eight"	American, "8 bits to dollar"

Time (Pennick, 1989):
8 days = old pagan week

We suspect that these are just a few of many examples of measurement systems which were based on the number eight. A complete study of this field would examine the history of all cultures and systems of measurement that have become obsolete.

Eight-fold Path of Lord Buddha (Feuerstein, 1994)

Lord Buddha taught eight practices to assist the seeker of truth to gain enlightenment and eliminate ignorance and suffering: right view, right resolve, right speech, right conduct, right livelihood, right endeavor, right mindfulness, and right meditation. Interestingly, in Buddhism there are eight auspicious symbols associated with Buddha's spiritual dominion and thirty-two marks of divine order.

Eight Limbs of Yoga (Feuerstein, 1994)

This is an aspect of Vedic Literature based on the wisdom from the enlightened sage Patanjali. Patanjali's teachings, often described as the eight limbs of Yoga, provide the knowledge for spiritual liberation. The eight limbs include: moral observance, self-restraint, posture, breath control, sensory inhibition, concentration, meditation, and samadhi (pure consciousness).

Dances, sports and games:

Many traditional dances have eight dancers in the standard set. In sports there are the "eights" in rowing and the "figure eight" of ice-skating. Numerous games are based on the number eight (e.g., the eight ball in pool, chess, checkers, the Irish game Tawlwrdd in which a king is defended by eight pieces and eight swordsmen).

Myths and symbols:

This is an area of almost endless examples, research and possibilities. The following are a few illustrations:

- The Hawaiian Cross contains a symbol of the Flower of the Sun. Radiating from this image are eight solar deities that represent the living water of life (Arguelles and Arguelles, 1985).

- In Australia, the Tjuringa Stone, the symbol of the Dream Time, represents eternity. This symbol has eight cardinal points and is a map for understanding the origin of all things (Arguelles and Arguelles, 1985).

- Throughout ancient India, the Lotus Blossom is the principle archetypal symbol used in yantras. Its eight petals radiating outwards illustrate the creative power or the divine essence. The numbers eight and sixty-four are also found in many other yantras and symbols from the Vedic age.

- In folklore it is believed that during the earlier part of this century there were eight Tibetan families whose function was to guard the eight directions using rituals. A similar, but earlier myth, originates from Celtic Britain where eight noble families were designated to perform the same function (Pennick, 1989).

- In pre-Christian Europe, the number eight is found throughout sagas and myths. It is usually associated with the central point of a story or the idea of fixity. For instance, northern myths tell of a trickster god called Loki who lived in the center of the Earth for eight winters, a hero Siegfried who served King Gibich for eight years, and Swan-Maidens who lived with Wayland the Smith and his brothers for eight winters. In another saga, a Druid named Cathbad is believed to have had eight disciples (Pennick, 1989).

A second avenue for future research is to use the foundational wisdom of the Constitution of the Universe to derive all other details on the mechanics of creation. In the second half of this book, we provided a prime example of how to derive the primal mechanics for the genesis of the material universe from the Constitution of the Universe. Perhaps a more in-depth analysis of the numerical/phonetic structure of the hymns of Rig Veda will reveal how the rest of creation sequentially unfolds from the fundamental Constitution of the Universe. As the reader may remember, our original understanding of the Constitution of the Universe was based on the literary structure of the very beginning hymns of the Veda. Thus, it is possible that an analysis of the numerical framework of the succeeding hymns of Rig Veda could provide additional secrets in the restoration of ancient knowledge.

Another tremendous source of universal wisdom is the surviving literature from the ancient world. In the Vedic Literature alone an estimated two to three million ancient texts lie scattered throughout the world. Out of these, only between 35,000 and 40,000 texts have been systematically studied with translations and attempted interpretations. The possibility for research and rediscovery of ancient wisdom is enormous. We are convinced that Vedic Literature combined with other texts from the ancient world will occupy the research of scholars and scientists for many decades, if not hundreds of years, to come.

REFERENCES

Adam, J. (1985). *The Nuptial Number.* (1st Edition published in 1891.) Northamptonshire, Great Britain: Thorsons Publishers Limited.

Amber, R. and Babey-Brooke, A.M. (1966). *The Pulse in Occident and Orient: Its Philosophy and Practice in Holistic Diagnosis and Treatment.* New York, NY: Aurora Press.

Andrews, W.S. (1960). *Magic Squares and Cubes.* New York, NY: Dover Publications, Inc.

Anton, F. (1978). *Art of the Maya.* London, England: Thames and Hudson.

Arguelles, J.A. (1984). *Earth Ascending: An Illustrated Treatise on the Law Governing Whole Systems.* Boulder, CO: Shambhala Publications, Inc.

Arguelles, J.A. (1987). *The Mayan Factor: Path Beyond Technology.* Santa Fe, NM: Bear and Co.

Arguelles, J.A. and Arguelles, M. (1985). *Mandala.* Boston, MA: Shambhala Publications, Inc.

Asimov, I. (1963). *The Human Brain: Its Capacities and Functions.* New York, NY: The New American Library, Inc.

Asimov, I. (1965). *A Short History of Chemistry.* Garden City, NY: Anchor Books, Doubleday & Co., Inc.

Azaroff, L.V. (1960). *Introduction to Solids.* New York, NY: McGraw-Hill Book Co., Inc.

Backus, J. (1969). *The Acoustical Foundations of Music.* New York, NY: W.W. Norton and Co., Inc.

Birren, F. (1963). *Color: A Survey in Words and Pictures, From Ancient Mysticism to Modern Science.* New Hyde Park, NY: University Books Inc.

Birren, F. (1969a). *A Basic Treatise on the Color System of Albert H. Munsell: A Grammar of Color.* New York, NY: Van Nostrand Reinhold Co.

Birren, F. (1969b). *A Basic Treatise on the Color System of Wilhelm Ostwald: The Color Primer.* New York, NY: Van Nostrand Reinhold Co.

Buckman, S.S. (1909). *An Octaval Instead of a Decimal System: An Essay to Show the Advantages of an Eight-figure, and the Disadvantages of a Ten-figure Notation for Money, Weights and Measures.* London, England: Simpkin, Marshall, Hamilton, Kent and Co., Ltd.

Chia, M. and Chia, M. (1989). *Fusion of the Five Elements I: Basic and Advanced Meditations for Transforming Negative Emotions.* Huntington, NY: Healing Tao Books.

Chopra, D. (1990). *Perfect Health: The Complete Mind/Body Guide.* New York, NY: Harmony Books.

Coe, M.D. (1971). *The Maya.* New York, NY: Praeger Publishers.

Connelly, D.M. (1979). *Traditional Acupuncture: The Law of the Five Elements.* Columbia, MD: The Centre for Traditional Acupuncture Inc.

da Vinci, L. (1802). *A Treatise on Painting.* London, England: Architectural Library, High Holborn.

Das, P.K. (1989). *The Secrets of Vastu.* Sikh Village, Secunderabad, India: Udayalakshmi Publications.

De Paz, M. and De Paz, M. (1993). *The Mayan Calendar: The Infinite Path of Time.* Guatemala City, Guatemala: Gran Jaguar 2.

Dillbeck, M.C. (1989). "Experience of the Ved: Realization of the Cosmic Psyche by Direct Perception: Opening Individual Awareness to the Self-interacting Dynamics of Consciousness." *Modern Science and Vedic Science,* 3(2): 116-152.

Doczi, G. (1981). *The Power of Limits: Proportional Harmonies in Nature, Art and Architecture.* Boulder, CO: Shambhala Publications, Inc.

Efron, A. (1941). *The Sacred Tree Script: The Esoteric Foundation of Plato's Wisdom.* New Haven, CT: The Tuttle, Morehouse and Taylor Co.

Falkener, E. (1961). *Games Ancient and Oriental and How to Play Them: Being the Games of the Ancient Egyptians the Hiera Gramme of the Greeks, the Ludus Latrunculorum of the Romans and the Oriental Games of Chess, Draughts, Backgammon and Magic Squares.* New York, NY: Dover Publications.

Feuerstein, G. (1994). *Spirituality by the Numbers.* New York, NY: G.P. Putnam's Sons.

Fortune, D. (1984). *Mystical Qabalah.* York Beach, ME: Samuel Weiser, Inc.

Frawley, D. (1986). *Hymns from the Golden Age: Selected Hymns from the Rig Veda with the Yogic Interpretation.* New Delhi, India: Motilal Banarsidass Publishers.

Frawley, D. (1992). *From the River of Heaven: Hindu and Vedic Knowledge for the Modern Age.* New Delhi, India: Motilal Banarsidass Publishers.

Frawley, D. (1993). *Gods, Sages and Kings: Vedic Secrets of Ancient Civilization.* New Delhi, India: Motilal Banarsidass Publishers.

Frawley, D. (1994). *Wisdom of the Ancient Seers: Mantras of the Rig Veda.* New Delhi, India: Motilal Banarsidass Publishers.

Frawley, D. and Lad, V. (1986). *The Yoga of Herbs: An Ayurvedic Guide to Herbal Medicine.* Twin Lakes, WI: Lotus Press.

Frissell, B. (1994). *Nothing in This Book Is True, But It's Exactly How Things Are: The Esoteric Meaning of the Monuments on Mars.* Berkeley, CA: Frog, Ltd.

Fuller, R.B. (1975). *Synergetics: Explorations in the Geometry of Thinking.* New York, NY: MacMillan Publishing Co., Inc.

Gescheider, G.A. and Niblette, R.K. (1967). *Cross-modality Masking for Touch and Hearing.* Journal of Experimental Psychology, 74(3): 313-320.

Grout, D.J. (1973). *A History of Western Music.* New York, NY: W.W. Norton and Co., Inc.

Guthrie, K.S. (1987). *The Pythagorean Sourcebook and Library: An Anthology of Ancient Writings which Related to Pythagoras and Pythagorean Philosophy.* Grand Rapids, MI: Phanes Press.

Hagelin, J.S. (1987). "Is Consciousness the Unified Field? A Field Theorist's Perspective." *Modern Science and Vedic Science,* 1(1): 28-87.

Hagelin, J.S. (1989). "Restructuring Physics from its Foundations in Light of Maharishi's Vedic Science." *Modern Science and Vedic Science,* 3(1): 3-72.

Hagelin, J.S. (1992). "The Constitution of the Universe: An Introduction". *MIU Video Magazine,* Maharishi International University: Volume 5, Tape 4.

Halevi, Z.S. (1972). *Tree of Life: An introduction to the Cabala.* London, England: Rider and Co.

Halevi, Z.S. (1974). *ADAM and the Kabbalistic Tree.* York Beach, ME: Samuel Weiser, Inc.

Halevi, Z.S. (1979). *Kabbalah: Tradition of Hidden Knowledge.* New York, NY: Thames and Hudson.

Hanayama, S. (1960). *A History of Japanese Buddhism*. Tokyo, Japan: The CIIB Press.

Helmholtz, H.L.F. (1954) *On the Sensations of Tone: As a Physiological Basis for the Theory of Music, 4th Edition*. New York, NY: Dover Publications, Inc.

Heninger, S.K. (1974). *Touches of Sweet Harmony: Pythagorean Cosmology and Renaissance Poetics*. San Marino, CA: The Huntington Library.

Hindemith, P. (1945). *The Craft of Musical Composition: Book I – Theoretical Part*. New York, NY: Associated Music Publishers.

Holden, A. and Morrison, P. (1982). *Crystals and Crystal Growing*. Cambridge, MA: The MIT Press.

Holtzman, S.R. (1994). *Digital Mantras*. Cambridge, MA: The MIT Press.

Hubble, E (1958). *The Realm of the Nebulae*. New York, NY: Dover Publications, Inc.

Huntley, H.E. (1970). *The Divine Proportion: A Study in Mathematical Beauty*. New York, NY: Dover Publications.

Hurtak, J.J. (1977). *The Book of Knowledge: The Keys of Enoch*. USA: The Academy for the Future of Science.

Jacobson, E.G. (1937). *The Science of Color: A Summary of the Theories of Dr. Wilhelm Ostwald*. St. Louis, MO: Barnes-Crosby Co.

Jeans, Sir J. (1968). *Science and Music*. New York: NY: Dover Publications, Inc.

Kak, S.C. (1993a). "Planetary Periods from the Rig Vedic Code." *The Mankind Quarterly,* 33(4): 433.

Kak, S.C. (1993b). *Astronomy of the Vedic Altars*. Vistas in Astronomy, 36(1): 117-140.

Kak, S.C. (1994). *The Astronomical Code of the Rig Veda*. Current Science, 66(4): 323.

Kak, S.C. and Frawley, D. (1992). *Further Observations on the Rig Vedic Code*. The Mankind Quarterly 33(2): 163.

Kaplan, R.A. (1981). *The Living Torah: A New Translation Based on Traditional Jewish Sources*. New York, NY: Maznaim Publishing Corp.

Kelley, S. (1993). "A Symposium on the Constitution of the Universe: World-Sheet Fermions and the Syllables of the Ved." *MIU Video Magazine,* Maharishi International University: Volume 6, Tape 3.

Kramrisch, S. (1976). *The Hindu Temple.* New Delhi, India: Motilal Banarsidass.

Lad, V. (1984). *Ayurveda: The Science of Self-Healing.* Twin Lakes, WI: Lotus Press.

Lai, T.C. (1972). *The Eight Immortals.* Lock Road, Kowloon, Hong Kong: Swindon Book Co.

Lao, T. (1961). *Lao Tzu / Tao Teh Ching.* New York, NY: St. John's University Press.

Lawlor, R. (1982). *Sacred Geometry: Philosophy and Practice.* New York, NY: Crossroad.

Legge, J. (1963). *The Sacred Books of China: The I Ching, 2nd Edition.* New York, NY: Dover Publications, Inc.

Lide, D.R. (1990). *CRC Handbook of Chemistry and Physics.* Boston, MA: CRC Press.

LIFE *Magazine* (1944). "Color: As the Response of Vision to Wave Lengths of Light." LIFE *Magazine,* July 3, Volume 17.

MacCurdy, E. (1941). *The Notebooks of Leonardo da Vinci.* New York, NY: Garden City Publishing Co., Inc.

Maharishi Mahesh Yogi (1967). *Maharishi Mahesh Yogi on the Bhagavad-Gita: A New Translation and Commentary; Chapters 1-6.* Baltimore, MD: Penguin.

Maharishi Mahesh Yogi (1986). *Life Supported by Natural Law: Lectures by His Holiness Maharishi Mahesh Yogi.* Washington, DC: Age of Enlightenment Press.

Maharishi Mahesh Yogi (1992). *Maharishi's Absolute Theory of Government: Automation in Administration.* Fairfield, IA: MIU Press.

Maharishi Mahesh Yogi (1994). *Vedic Knowledge for Everyone.* Vlodrop, Holland: Maharishi Vedic University Press.

Mantle, J. (1978). *Leonardo da Vinci: Anatomical Drawings.* Spain: Miller Graphics, Crown Publishers, Inc.

Marks, L.E. (1978). *The Unity of the Senses: Interrelations Among the Modalities.* New York, NY: Academic Press.

Melchizedek, D. (1992). "Flower of Life Workshop." Dallas, TX.

Michell, J. (1972). *City of Revelation: On the Proportions and Symbolic Numbers of the Cosmic Temple.* New York, NY: Ballantine Books.

Moore, K.L. and Persaud, T.V.N. (1993). *Before We Are Born: Essentials of Embryology and Birth Defects.* Philadelphia, PA: W.B. Saunders Co.

Morley, S.G. and Brainerd, G.W. (1983). *The Ancient Maya.* Stanford, CA: Stanford University Press.

Murray, H.J.R. (1962). *A History of Chess.* London, England: University Press.

Nader, T. (1995). *Human Physiology – Expression of Veda and the Vedic Literature.* Vlodrop, The Netherlands: Maharishi Vedic University Press.

Nickerson, D. (1946). "The Munsell Color System." *Illuminating Engineering* XLI (7): 549-560.

Nylan, M. (1994). *The Elemental Changes: The Ancient Chinese Companion to the I Ching.* Albany, NY: State University of New York Press.

Ohno, S. and Jabara, M. (1985). "Repeats of Base Oligomers ($N = 3n \pm 1$ or 2) as Immortal Coding Sequences of the Primeval World: Construction of Coding Sequences is Based upon the Principle of Musical Composition." *Chemica Scripta* 26B: 43-49.

Ohno, S. and Ohno, M. (1986). "The all Pervasive Principle of Repetitious Recurrence Governs not only Coding Sequence Construction but also Human Endeavor in Musical Composition." *Immunogenetics* 24: 71-78.

Parasara, M. – translation by R. Santhanam – (1984). *Brihat Parasara Hora Sastra of Maharishi Parasara.* New Delhi, India: Ranjan Publications.

Patel, S.V. (1992). *Sacred Geometry in Chess and the Design of the Hindu Temple.* Manhattan, KS: College of Architecture and Design, Kansas State University.

Pennick, N. (1989). *Secret Games of the Gods: Ancient Ritual Systems in Board Games.* York Beach, ME: Samuel Wiser, Inc.

Ponce, C. (1973). *Kabbalah.* USA: Theosophical Publishing House.

Reynal and Co. (1967). *Leonardo da Vinci.* Japan: Dainoppon Printing Co., Ltd.

Rossbach, S. (1987). *Interior Design with Feng Shui.* New York, NY: E.P. Dutton.

Scholem, G.G. (1965). *On the Kabbalah and its Symbolism.* New York, NY: Schoken Books.

Scholem, G.G. (1978). *Kabbalah*. New York, NY: Meridian.

Schonberger, M. – translation by D.Q. Stephenson – (1992). *The I Ching and the Genetic Code*. Santa Fe, NM: Aurora Press.

Settegast, M. (1987). *Plato Prehistorian: 10000 to 5000 BC in Myth and Archaeology*. Cambridge, MA: The Rotenberg Press.

Sharma, H. (1993). *Freedom from Disease*. Toronto, Ontario: Veda Publishing, Inc.

Sharma, R.K. and Dash, B. (translators; 1988). *Caraka Samhita*. Varanasi, India: Chowkhamba Sanskrit Studies.

Shearer, T. (1975). *Beneath the Moon and Under the Sun: A Poetic Re-appraisal of the Sacred Calendar and the Prophecies of Ancient Mexico*. Albuquerque, NM: Sun Publishing Co.

Sheinkin, D. (1986). *Path of the Kabbalah*. New York, NY: Paragon House.

Smith, C.P.W. and Williams, P.L. (1984). *Basic Human Embryology*. London, England: Pitman Publishing Limited.

Spence, A.P. and Mason, E.B. (1992). *Human Anatomy and Physiology*. New York, NY: West Publishing Co.

Starr, C. and Taggart, R. (1984). *Biology: The Unity and Diversity of Life, 3rd Edition*. Belmont, CA: Wadsworth Publishing Co.

Stent, G.S. (1969). *The Coming of the Golden Age: A View of the End of Progress*. Garden City, NY: Natural History Press.

Szekely, E.B. (1957). *The Teachings of the Essenes from Enoch to the Dead Sea Scrolls*. San Diego, CA: Academy Books.

Szekely, E.B. (1974). *The Essene Teachings of Zarathustra*. San Diego, CA: Academy Books.

Szekely, E.B. (1976). *The Essene Book of Asha: Journey to the Cosmic Ocean*. San Diego, CA: Academy Books.

Szekely, E.B. (1990). *The Zend Avesta of Zarathustra*. USA: International Biogenic Society.

Szekely, E.B. (1994). *Archeosophy, A New Science*. USA: International Biogenic Society.

Taylor, T. (1972). *The Theoretic Arithmetic of the Pythagoreans*. York Beech, ME: Samuel Weiser, Inc.

Todhunter, I. (1948). *The Elements of Euclid*. New York, NY: E.P. Dutton and Co., Inc.

Torrey, R.A., et. al. (Eds.; 1990). *The Fundamentals: The Famous Sourcebook of Foundational Biblical Truths.* Grand Rapids, MI: Kregal Publications.

Twersky, I (1972). *A Maimonides Reader.* New York, NY: Behrman House, Inc.

Vatsyayana (1992). *The Kamasutra of Vatsyayana.* New Delhi, India: Lustre Press Put Ltd.

Wallace, R.K. (1986). *The Maharishi Technology of the Unified Field: The Neurophysiology of Enlightenment.* Fairfield, IA: MIU Neuroscience Press.

Wallace, R.K. (1993). *The Physiology of Consciousness.* Fairfield, IA: MIU Press and Institute of Science, Technology and Public Policy.

Wallace, R.K., Fagan, J.B. and Pasco, D.S. (1988). "Vedic Physiology." *Modern Science and Vedic Science,* 2(1): 2-59.

Wilhelm, R. (1979). *Lectures on the I Ching.* Princeton and London: Princeton University Press.

Wilhelm, R. and Baynes C.F. (1967). *The I Ching or Book of Changes, 3rd Edition.* Princeton: Princeton University Press.

Wilhelm, R. and Jung, C.J. (1962). *The Secret of the Golden Flower: The Chinese Book of Life.* New York, NY: Harcourt, Brace and World, Inc.

Wood, E. A. (1964). *Crystals and Light: An Introduction to Optical Crystallography.* New Jersey: D. Van Nostrand Co., Inc.

Yan, J.F (1991). *DNA and the I Ching: The Tao of Life.* Berkeley, CA: North Atlantic Books.

Zumdahl, S.S. (1989). *Chemistry, 2nd Edition.* Lexington, MA: D.C. Heath and Co.

ADDITIONAL TITLES
BY SUNSTAR PUBLISHING LTD.

The Name Book by Pierre Le Rouzic
ISBN 0-9638502-1-0 $15.95
Numerology/Philosophy. International bestseller. Over 9,000 names with stunningly accurate descriptions of character and personality. How the sound of your name effects who you grow up to be.

Every Day A Miracle Happens by Rodney Charles
ISBN 0-9638502-0-2 $17.95
Religious bestseller. 365 stories of miracles, both modern and historic, each associated with a day of the year. Universal calendar. Western religion.

Of War & Weddings by Jerry Yellin
ISBN 0-9638502-5-3 $17.95
History/Religion. A moving and compelling autobiography of bitter wartime enemies who found peace through their children's marriage. Japanese history and religion.

Your Star Child by Mary Mayhew
ISBN 0-9638502-2-9 $16.95
East/West philosophy. Combines Eastern philosophy with the birthing techniques of modern medicine, from preconception to parenting young adults.

Lighter Than Air by Rodney Charles and Anna Jordan
ISBN 0-9638502-7-X $14.95
East/West philosophy. Historic accounts of saints, sages and holy people who possessed the ability of unaided human flight.

Bringing Home the Sushi by Mark Meers
ISBN 1-887472-05-3 $21.95
Japanese philosophy/culture. Adventurous account of of an American businessman and his family living in '90s Japan.

Miracle of Names by Clayne Conings
    ISBN 1-887472-03-7 $13.95
Numerology/Eastern philosophy. Educational and enlightening – discover
the hidden meanings and potential of names through numerology.

Voice for the Planet by Anna Maria Gallo
    ISBN 1-887472-00-2 $10.95
Religion/Ecology. This book explores the ecological practicality of native
American practices.

Making $$$ at Home by Darla Sims
    ISBN 1-887472-02-9 $25.00
Reference. Labor-saving directory that guides you through the process of
making contacts to create a business at home.

Gabriel & the Remarkable Pebbles by Carol Hovin
    ISBN 1-887472-06-1 $12.95
Children/Ecology. A lighthearted, easy-to-read fable that educates children
in understanding ecological balances.

Searching for Camelot by Edith Thomas
    ISBN 1-887472-08-8 $12.95
East/West philosophy. Short, easy-to-read, autobiographical adventure full
of inspirational life lessons.

The Revelations of Ho by Dr. James Weldon
    ISBN 1-887472-09-6 $17.95
Eastern philosophy. A vivid and detailed account of the path of a modern-
day seeker of enlightenment.

The Formula by Dr. Vernon Sylvest
    ISBN 1-887472-10-X $21.95
Eastern philosophy/Medical research. This book demystifies the gap
between medicine and mysticism, offering a ground breaking perspec-
tive on health as seen through the eyes of an eminent pathologist.

Jewel of the Lotus by Bodhi Avinasha
    ISBN 1-887472-11-8 $15.95
Eastern philosophy. Tantric Path to higher consciousness. Learn to
increase your energy level, heal and rejuvenate yourself through devo-
tional relationships.

Elementary, My Dear by Tree Stevens
  ISBN 1-887472-12-6 $17.95
Cooking/Health. Step-by-step, health-conscious cookbook for the beginner. Includes hundreds of time-saving menus.

Directory of New Age & Alternative Publications by Darla Sims
  ISBN 1-887472-18-5 $23.95
Reference. Comprehensive listing of publications, events, organizations arranged alphabetically, by category and by location.

Educating Your Star Child by Ed & Mary Mayhew
  ISBN 1-887472-17-7 $16.95
East/West philosophy. How to parent children to be smarter, wiser and happier, using internationally acclaimed mind-body intelligence techniques.

How to be Totally Unhappy in a Peaceful World by Gil Friedman
  ISBN 1-887472-13-4 $11.95
Humor/Self-help. Everything you ever wanted to know about being unhappy: A complete manual with rules, exercises, a midterm and final exam. *(Paper.)*

No Justice by Chris Raymondo
  ISBN 1-887472-14-2 $23.95
Adventure. Based on a true story, this adventure novel provides behind the scenes insight into CIA and drug cartel operations. One of the best suspense novels of the '90s. *(Cloth.)*

On Wings of Light by Ronna Herman
  ISBN 1-887472-19-3 $19.95
New Age. Ronna Herman documents the profoundly moving and inspirational messages for her beloved Archangel Michael.

The Global Oracle by Edward Tarabilda & Doug Grimes
  ISBN 1-887472-22-3 $17.95
East/West philosophy. A guide to the study of archetypes, with an excellent introduction to holistic living. Use this remarkable oracle for meditation, play or an aid in decision making.

Destiny by Sylvia Clute
ISBN 1-887472-21-5 $21.95
East/West philosophy. A brilliant metaphysical mystery novel (with the ghost of George Washington) based on A Course In Miracles.

The Husband's Manual by A. & T. Murphy
ISBN 0-9632336-4-5 $9.00
Self-help/Men's issues. At last! Instructions for men on what to do and when to do it. The Husband's Manual can help a man create a satisfying, successful marriage – one he can take pride in, not just be resigned to.

Cosmic Perspective by Harold Allen
ISBN 1-887472-23-1 $21.95
Science/Eastern philosophy. Eminent cosmologist Harold Allen disproves the "Big Bang Theory" and paves the way for the new era of "Consciousness Theory."

Twin Galaxies Pinball Book of World Records by Walter Day
ISBN 1-887472-25-8 $12.95
Reference. The official reference book for all Video Game and Pinball Players – this book coordinates an international schedule of tournaments that players can compete in to gain entrance into this record book.

How to Have a Meaningful Relationship with Your Computer
by Sandy Berger
ISBN 1-887472-36-3 $18.95
Computer/Self-help. A simple yet amusing guide to buying and using a computer, for beginners as well as those who need a little more encouragement.

The Face on Mars by Harold Allen
ISBN 1-887472-27-4 $14.95
Science/Fiction. A metaphysical/scientific novel based on the NASA space expedition to Mars.

The Spiritual Warrior by Shakura Rei
ISBN 1-887472-28-2 $17.95
Eastern philosophy. An exposition of the spiritual techniques and practices of Eastern Philosophy.

The Pillar of Celestial Fire by Robert Cox
    ISBN 1-887472-30-4 $18.95
Eastern philosophy. The ancient cycles of time, the sacred alchemical science and the new golden age.

The Tenth Man by Wei Wu Wei
    ISBN 1-887472-31-2 $15.95
Eastern philosophy. Discourses on Vedanta – the final stroke of enlightenment.

Open Secret by Wei Wu Wei
    ISBN 1-887472-32-0 $14.95
Eastern philosophy. Discourses on Vedanta – the final stroke of enlightenment.

All Else is Bondage by Wei Wu Wei
    ISBN 1-887472-34-7 $16.95
Eastern philosophy. Discourses on Vedanta – the final stroke of enlightenment.

Publishing and Distributing your Book is this Simple ...

1. Send us your completed manuscript.
2. We'll review it. Then after acceptance we'll:
   - Register your book with The Library of Congress, Books in Print, and acquire International Standard Book Numbers, including UPC Bar Codes.
   - Design and print your book cover.
   - Format and produce 150 review copies.
   - Deliver review copies (with sales aids) to 20 of the nation's leading distributors and 50 major newspaper, magazine, television and radio book reviewers in the USA and Canada.
   - Organize author interviews and book reviews.
3. Once we have generated pre-orders for 1,000 books, New Author Enterprises will enter into an exclusive publishing contract offering up to 50% profit-sharing terms with the author.

People are Talking about Us ...

"I recommend New Author Enterprises to any new author. The start-up cost to publish my book exceeded $20,000 – making it nearly impossible for me to do it on my own. New Author Enterprises' ingenious marketing ideas and their network of distributors allowed me to reach my goals for less than $4,000. Once my book reached the distributors and orders started coming in, New Author Enterprises handled everything – financing, printing, fulfillment, marketing – and I earned more than I could have with any other publisher."

– Rodney Charles, best-selling author of
Every Day A Miracle Happens

Publish It Now!

204 North Court St., Fairfield IA 52556 • (800) 532-4734
http://www.newagepage.com